ALGEBRAIC GRAPH THEORY

ALGEBRAIC GRAPH THEORY

Second Edition

NORMAN BIGGS
London School of Economics

CAMBRIDGE
UNIVERSITY PRESS

CAMBRIDGE UNIVERSITY PRESS

Cambridge, New York, Melbourne, Madrid, Cape Town, Singapore,
São Paulo, Delhi, Dubai, Tokyo, Mexico City

Cambridge University Press
The Edinburgh Building, Cambridge CB2 8RU, UK

Published in the United States of America by
Cambridge University Press, New York

www.cambridge.org
Information on this title: www.cambridge.org/9780521458979

First published 1974
Second edition 1993
Reprinted 1996

A catalogue record for this publication is available from the British Library

Library of Congress cataloguing in publication data available

ISBN 978-0-521-20335-7 Hardback
ISBN 978-0-521-45897-9 Paperback

Contents

Preface   vii

1 Introduction   1

PART ONE – LINEAR ALGEBRA IN GRAPH THEORY

2 The spectrum of a graph   7
3 Regular graphs and line graphs   14
4 Cycles and cuts   23
5 Spanning trees and associated structures   31
6 The tree-number   38
7 Deteminant expansions   44
8 Vertex-partitions and the spectrum   52

PART TWO – COLOURING PROBLEMS

9 The chromatic polynomial   63
10 Subgraph expansions   73
11 The multiplicative expansion   81
12 The induced subgraph expansion   89
13 The Tutte polynomial   97
14 Chromatic polynomials and spanning trees   106

PART THREE – SYMMETRY AND REGULARITY

15 Automorphisms of graphs   115
16 Vertex-transitive graphs   122
17 Symmetric graphs   130

18 Symmetric graphs of degree three   138
19 The covering-graph construction   149
20 Distance-transitive graphs   155
21 Feasibility of intersection arrays   164
22 Imprimitivity   173
23 Minimal regular graphs with given girth   180

References   191
Index   202

Preface

This book is a substantially enlarged version of the Cambridge Tract with the same title published in 1974. There are two major changes.

- The main text has been thoroughly revised in order to clarify the exposition, and to bring the notation into line with current practice. In the course of revision it was a pleasant surprise to find that the original text remained a fairly good introduction to the subject, both in outline and in detail. For this reason I have resisted the temptation to reorganise the material in order to make the book rather more like a standard textbook.

- Many *Additional Results* are now included at the end of each chapter. These replace the rather patchy selection in the old version, and they are intended to cover most of the major advances in the last twenty years. It is hoped that the combination of the revised text and the additional results will render the book of service to a wide range of readers.

I am grateful to all those people who have helped by commenting upon the old version and the draft of the new one. Particular thanks are due to Peter Rowlinson, Tony Gardiner, Ian Anderson, Robin Wilson, and Graham Brightwell. On the practical side, I thank Alison Adcock, who prepared a TEX version of the old book, and David Tranah of Cambridge University Press, who has been constant in his support.

Norman Biggs   March 1993

Preface

1

Introduction to algebraic graph theory

About the book

This book is concerned with the use of algebraic techniques in the study of graphs. The aim is to translate properties of graphs into algebraic properties and then, using the results and methods of algebra, to deduce theorems about graphs.

It is fortunate that the basic terminology of graph theory has now become part of the vocabulary of most people who have a serious interest in studying mathematics at this level. A few basic definitions are gathered together at the end of this chapter for the sake of convenience and standardization. Brief explanations of other graph-theoretical terms are included as they are needed. A small number of concepts from matrix theory, permutation-group theory, and other areas of mathematics, are used, and these are also accompanied by a brief explanation.

The literature of algebraic graph theory itself has grown enormously since 1974, when the original version of this book was published. Literally thousands of research papers have appeared, and the most relevant ones are cited here, both in the main text and in the Additional Results at the end of each chapter. But no attempt has been made to provide a complete bibliography, partly because there are now several books dealing with aspects of this subject. In particular there are two books which contain massive quantities of information, and on which it is convenient to rely for 'amplification and exemplification' of the main results discussed here.

These are:

Spectra of Graphs: D.M. Cvetković, M. Doob, and H. Sachs, Academic Press (New York) 1980.

Distance-Regular Graphs: A.E. Brouwer, A.M. Cohen, and A. Neumaier, Springer-Verlag (Berlin) 1989.

References to these two books are given in the form [CvDS, p. 777], and [BCN, p. 888].

C.D. Godsil's recent book *Algebraic Combinatorics* (Chapman and Hall, 1993) arrived too late to be quoted as reference. It is in many ways complementary to this book, since it covers several of the same topics from a different point of view. Finally, the long-awaited *Handbook of Combinatorics* will contain authoritative accounts of many subjects discussed in these pages.

Outline of the book

The book is in three parts, each divided into a number of short chapters. The first part deals with the applications of linear algebra and matrix theory to the study of graphs. We begin by introducing the adjacency matrix of a graph; this matrix completely determines the graph, and its spectral properties are shown to be related to properties of the graph. For example, if a graph is regular, then the eigenvalues of its adjacency matrix are bounded in absolute value by the degree of the graph. In the case of a line graph, there is a strong lower bound for the eigenvalues. Another matrix which completely describes a graph is the incidence matrix of the graph. This matrix represents a linear mapping which determines the homology of the graph. The problem of choosing a basis for the homology of a graph is just that of finding a fundamental system of cycles, and this problem is solved by using a spanning tree. At the same time we study cuts in the graph. These ideas are then applied to the systematic solution of network equations, a topic which supplied the stimulus for the original theoretical development. We then investigate formulae for the number of spanning trees in a graph, and results which are derived from the expansion of determinants. These expansions illuminate the relationship between a graph and the characteristic polynomial of its adjacency matrix. The first part ends with a discussion of how spectral techniques can be used in problems involving partitions of the vertex-set, such as the vertex-colouring problem.

The second part of the book deals with the colouring problem from a different point of view. The algebraic technique for counting the colourings of a graph is founded on a polynomial known as the chromatic polynomial. We first discuss some simple ways of calculating this polynomial, and show how these can be applied in several important cases. Many important properties of the chromatic polynomial of a graph stem from its connection with the family of subgraphs of the graph, and we show how the chromatic polynomial can be expanded in terms of subgraphs. From the first (additive) expansion another (multiplicative) expansion can be derived, and the latter depends upon a very restricted class of subgraphs. This leads to efficient methods for approximating the chromatic polynomials of large graphs. A completely different kind of expansion relates the chromatic polynomial to the spanning trees of a graph; this expansion has several remarkable features and leads to new ways of looking at the colouring problems, and some new properties of chromatic polynomials.

The third part of the book is concerned with symmetry and regularity properties. A symmetry property of a graph is related to the existence of automorphisms – that is, permutations of the vertices which preserve adjacency. A regularity property is defined in purely numerical terms. Consequently, symmetry properties induce regularity properties, but the converse is not necessarily true. We first study the elementary properties of automorphisms, and explain the connection between the automorphisms of a graph and the eigenvalues of its adjacency matrix. We then introduce a hierarchy of symmetry conditions which can be imposed on a graph, and proceed to investigate their consequences. The condition that all vertices be alike (under the action of the group of automorphisms) turns out to be rather a weak one, but a slight strengthening of it leads to highly non-trivial conclusions. In fact, under certain conditions, there is an absolute bound to the level of symmetry which a graph can possess. A strong symmetry property, called distance-transitivity, and the consequent regularity property, called distance-regularity, are then introduced. We return to the methods of linear algebra to derive numerical constraints upon the existence of graphs with these properties. Finally, these constraints are applied to the problem of finding minimal regular graphs whose degree and girth are given.

Basic definitions and notation

Formally, a *general graph* Γ consists of three things: a set $V\Gamma$, a set $E\Gamma$, and an incidence relation, that is, a subset of $V\Gamma \times E\Gamma$. An element of $V\Gamma$ is called a *vertex*, an element of $E\Gamma$ is called an *edge*, and the incidence relation is required to be such that an edge is incident with either one vertex (in which case it is a *loop*) or two vertices. If every edge is incident with two vertices, and no two edges are incident with the same pair of vertices, then we say that Γ is a *strict graph* or briefly, a *graph*. In this case, $E\Gamma$ can be regarded as a subset of the set of unordered pairs of vertices. We shall deal mainly with graphs (that is, strict graphs), except in Part Two, where it is sometimes essential to consider general graphs.

If v and w are vertices of a graph Γ, and $e = \{v, w\}$ is an edge of Γ, then we say that e *joins* v and w, and that v and w are the *ends* of e. The number of edges of which v is an end is called the *degree* of v. A *subgraph* of Γ is constructed by taking a subset S of $E\Gamma$ together with all vertices incident in Γ with some edge belonging to S. An *induced subgraph* of Γ is obtained by taking a subset U of $V\Gamma$ together with all edges which are incident in Γ only with vertices belonging to U. In both cases the incidence relation in the subgraph is inherited from the incidence relation in Γ. We shall use the notation $\langle S \rangle_\Gamma$, $\langle U \rangle_\Gamma$ for these subgraphs, and usually, when the context is clear, the subscript Γ will be omitted.

PART ONE

Linear algebra in graph theory

2

The spectrum of a graph

We begin by defining a matrix which will play an important role in many parts of this book. Suppose that Γ is a graph whose vertex-set $V\Gamma$ is the set $\{v_1, v_2, ..., v_n\}$, and consider $E\Gamma$ as a set of unordered pairs of elements of $V\Gamma$. If $\{v_i, v_j\}$ is in $E\Gamma$, then we say that v_i and v_j are *adjacent*.

Definition 2.1 The *adjacency matrix* of Γ is the $n \times n$ matrix $\mathbf{A} = \mathbf{A}(\Gamma)$ whose entries a_{ij} are given by

$$a_{ij} = \begin{cases} 1, & \text{if } v_i \text{ and } v_j \text{ are adjacent;} \\ 0, & \text{otherwise.} \end{cases}$$

For the sake of definiteness we consider $\mathbf{A}$ as a matrix over the complex field. Of course, it follows directly from the definition that $\mathbf{A}$ is a real symmetric matrix, and that the trace of $\mathbf{A}$ is zero. Since the rows and columns of $\mathbf{A}$ correspond to an arbitrary labelling of the vertices of Γ, it is clear that we shall be interested primarily in those properties of the adjacency matrix which are invariant under permutations of the rows and columns. Foremost among such properties are the spectral properties of $\mathbf{A}$.

Suppose that λ is an eigenvalue of $\mathbf{A}$. Then, since $\mathbf{A}$ is real and symmetric, it follows that λ is real, and the multiplicity of λ as a root of the equation $\det(\lambda\mathbf{I} - \mathbf{A}) = 0$ is equal to the dimension of the space of eigenvectors corresponding to λ.

Definition 2.2   The *spectrum* of a graph Γ is the set of numbers which are eigenvalues of $\mathbf{A}(\Gamma)$, together with their multiplicities. If the distinct eigenvalues of $\mathbf{A}(\Gamma)$ are $\lambda_0 > \lambda_1 > \ldots > \lambda_{s-1}$, and their multiplicities are $m(\lambda_0), m(\lambda_1), \ldots, m(\lambda_{s-1})$, then we shall write

$$\operatorname{Spec}\Gamma = \begin{pmatrix} \lambda_0 & \lambda_1 & \ldots & \lambda_{s-1} \\ m(\lambda_0) & m(\lambda_1) & \ldots & m(\lambda_{s-1}) \end{pmatrix}.$$

For example, the *complete graph* K_n is the graph with n vertices in which each distinct pair are adjacent. Thus the graph K_4 has adjacency matrix

$$\mathbf{A} = \begin{bmatrix} 0 & 1 & 1 & 1 \\ 1 & 0 & 1 & 1 \\ 1 & 1 & 0 & 1 \\ 1 & 1 & 1 & 0 \end{bmatrix},$$

and an easy calculation shows that the spectrum of K_4 is

$$\operatorname{Spec} K_4 = \begin{pmatrix} 3 & -1 \\ 1 & 3 \end{pmatrix}.$$

We shall usually refer to the eigenvalues of $\mathbf{A} = \mathbf{A}(\Gamma)$ as the *eigenvalues of* Γ. Also, the characteristic polynomial $\det(\lambda\mathbf{I} - \mathbf{A})$ will be referred to as the *characteristic polynomial of* Γ, and denoted by $\chi(\Gamma; \lambda)$. Let us suppose that the characteristic polynomial of Γ is

$$\chi(\Gamma; \lambda) = \lambda^n + c_1\lambda^{n-1} + c_2\lambda^{n-2} + c_3\lambda^{n-3} + \ldots + c_n.$$

In this form we know that $-c_1$ is the sum of the zeros, that is, the sum of the eigenvalues. This is also the trace of $\mathbf{A}$ which, as we have already noted, is zero. Thus $c_1 = 0$. More generally, it is proved in the theory of matrices that all the coefficients can be expressed in terms of the *principal minors* of $\mathbf{A}$, where a principal minor is the determinant of a submatrix obtained by taking a subset of the rows and the same subset of the columns. This leads to the following simple result.

Proposition 2.3   *The coefficients of the characteristic polynomial of a graph Γ satisfy:*
   (1) $c_1 = 0$;
   (2) $-c_2$ *is the number of edges of* Γ;
   (3) $-c_3$ *is twice the number of triangles in* Γ.

Proof   For each $i \in \{1, 2, \ldots, n\}$, the number $(-1)^i c_i$ is the sum of those principal minors of $\mathbf{A}$ which have i rows and columns. So we can argue as follows.
   (1) Since the diagonal elements of $\mathbf{A}$ are all zero, $c_1 = 0$.
   (2) A principal minor with two rows and columns, and which has a

non-zero entry, must be of the form

$$\begin{vmatrix} 0 & 1 \\ 1 & 0 \end{vmatrix}.$$

There is one such minor for each pair of adjacent vertices of Γ, and each has value -1. Hence $(-1)^2 c_2 = -|E\Gamma|$, giving the result.

(3) There are essentially three possibilities for non-trivial principal minors with three rows and columns:

$$\begin{vmatrix} 0 & 1 & 0 \\ 1 & 0 & 0 \\ 0 & 0 & 0 \end{vmatrix}, \quad \begin{vmatrix} 0 & 1 & 1 \\ 1 & 0 & 0 \\ 1 & 0 & 0 \end{vmatrix}, \quad \begin{vmatrix} 0 & 1 & 1 \\ 1 & 0 & 1 \\ 1 & 1 & 0 \end{vmatrix},$$

and, of these, the only non-zero one is the last (whose value is 2). This principal minor corresponds to three mutually adjacent vertices in Γ, and so we have the required description of c_3. ◻

These simple results indicate that the characteristic polynomial of a graph is an object of the kind we study in algebraic graph theory: it is an algebraic construction which contains graphical information. Proposition 2.3 is just a pointer, and we shall obtain a more comprehensive result on the coefficients of the characteristic polynomial in Chapter 7.

Suppose $\mathbf{A}$ is the adjacency matrix of a graph Γ. Then the set of polynomials in $\mathbf{A}$, with complex coefficients, forms an algebra under the usual matrix operations. This algebra has finite dimension as a complex vector space. Indeed, the Cayley–Hamilton theorem asserts that $\mathbf{A}$ satisfies its own characteristic equation, so the dimension is at most n, the number of vertices in Γ.

Definition 2.4 The *adjacency algebra* of a graph Γ is the algebra of polynomials in the adjacency matrix $\mathbf{A} = \mathbf{A}(\Gamma)$. We shall denote the adjacency algebra of Γ by $\mathcal{A}(\Gamma)$.

Since every element of the adjacency algebra is a linear combination of powers of $\mathbf{A}$, we can obtain results about $\mathcal{A}(\Gamma)$ from a study of these powers. We define a *walk* of length l in Γ, from v_i to v_j, to be a finite sequence of vertices of Γ,

$$v_i = u_0, u_1, ..., u_l = v_j,$$

such that u_{t-1} and u_t are adjacent for $1 \leq t \leq l$.

Lemma 2.5 *The number of walks of length l in Γ, from v_i to v_j, is the entry in position (i, j) of the matrix $\mathbf{A}^l$.*

Proof The result is true for $l = 0$ (since $\mathbf{A}^0 = \mathbf{I}$) and for $l = 1$ (since $\mathbf{A}^1 = \mathbf{A}$ is the adjacency matrix). Suppose that the result is true for $l = L$. The set of walks of length $L + 1$ from v_i to v_j is in bijective correspondence with the set of walks of length L from v_i to vertices v_h adjacent to v_j. Thus the number of such walks is

$$\sum_{\{v_h,v_j\}\in E\Gamma} (\mathbf{A}^L)_{ih} = \sum_{h=1}^{n}(\mathbf{A}^L)_{ih}a_{hj} = (\mathbf{A}^{L+1})_{ij}.$$

It follows that the number of walks of length $L+1$ joining v_i to v_j is $(\mathbf{A}^{L+1})_{ij}$. The general result follows by induction. □

A graph is said to be *connected* if each pair of vertices is joined by a walk. The number of edges traversed in the shortest walk joining v_i and v_j is called the *distance* in Γ between v_i and v_j and is denoted by $\partial(v_i, v_j)$. The maximum value of the distance function in a connected graph Γ is called the *diameter* of Γ.

Proposition 2.6 *Let Γ be a connected graph with adjacency algebra $\mathcal{A}(\Gamma)$ and diameter d. Then the dimension of $\mathcal{A}(\Gamma)$ is at least $d+1$.*

Proof Let x and y be vertices of Γ such that $\partial(x, y) = d$, and suppose that

$$x = w_0, w_1, \ldots, w_d = y$$

is a walk of length d. Then, for each $i \in \{1, 2, \ldots, d\}$, there is at least one walk of length i, but no shorter walk, joining w_0 to w_i. Consequently, $\mathbf{A}^i$ has a non-zero entry in a position where the corresponding entries of $\mathbf{I}, \mathbf{A}, \mathbf{A}^2, \ldots, \mathbf{A}^{i-1}$ are zero. It follows that $\mathbf{A}^i$ is not linearly dependent on $\{\mathbf{I}, \mathbf{A}, \ldots, \mathbf{A}^{i-1}\}$, and that $\{\mathbf{I}, \mathbf{A}, \ldots, \mathbf{A}^d\}$ is a linearly independent set in $\mathcal{A}(\Gamma)$. Since this set has $d+1$ members, the proposition is proved. □

There is a close connection between the adjacency algebra and the spectrum of Γ. If the adjacency matrix has s distinct eigenvalues then, since it is a real symmetric matrix, its minimum polynomial (the monic polynomial of least degree which annihilates it) has degree s. Consequently the dimension of the adjacency algebra is equal to s. Thus we have the following bound for the number of distinct eigenvalues.

Corollary 2.7 *A connected graph with diameter d has at least $d+1$ distinct eigenvalues.* □

One of the major topics of the last part of this book is the study of a class of 'highly regular' connected graphs which have the minimum number $d+1$ of distinct eigenvalues. In the following chapters we shall encounter several other examples of the link between structural regularity and the spectrum.

Notation The eigenvalues of a graph may be be listed in two ways: in strictly decreasing order of the distinct values, as in Definition 2.2, or in weakly decreasing order (with repeated values) $\lambda_0 \geq \lambda_1 \geq \ldots \geq \lambda_{n-1}$, where $n = |V\Gamma|$. We shall use either method, as appropriate.

Additional Results

2a *A reduction formula for* χ Suppose Γ is a graph with a vertex v_1 of degree 1, and let v_2 be the vertex adjacent to v_1. Let Γ_1 be the induced subgraph obtained by removing v_1, and Γ_{12} the induced subgraph obtained by removing $\{v_1, v_2\}$. Then

$$\chi(\Gamma; \lambda) = \lambda\chi(\Gamma_1; \lambda) - \chi(\Gamma_{12}; \lambda).$$

This formula can be used to calculate the characteristic polynomial of any tree, because a tree always has a vertex of degree 1. A more general reduction formula was found by Rowlinson (1987).

2b *The characteristic polynomial of a path*   Let P_n be the *path graph* with vertex-set $\{v_1, v_2, \ldots, v_n\}$ and edges $\{v_i, v_{i+1}\}$ $(1 \leq i \leq n-1)$. For $n \geq 3$ we have

$$\chi(P_n; \lambda) = \lambda\chi(P_{n-1}; \lambda) - \chi(P_{n-2}; \lambda).$$

Hence $\chi(P_n; \lambda) = U_n(\lambda/2)$, where U_n denotes the *Chebyshev polynomial of the second kind*.

2c *The spectrum of a bipartite graph*   A graph is *bipartite* if its vertex-set can be partitioned into two parts V_1 and V_2 such that each edge has one vertex in V_1 and one vertex in V_2. If we order the vertices so that those in V_1 come first, then the adjacency matrix of a bipartite graph takes the form

$$\mathbf{A} = \begin{bmatrix} \mathbf{0} & \mathbf{B} \\ \mathbf{B}^t & \mathbf{0} \end{bmatrix}.$$

If $\mathbf{x}$ is an eigenvector corresponding to the eigenvalue λ, and $\tilde{\mathbf{x}}$ is obtained from $\mathbf{x}$ by changing the signs of the entries corresponding to vertices in V_2, then $\tilde{\mathbf{x}}$ is an eigenvector corresponding to the eigenvalue $-\lambda$. It follows that the spectrum of a bipartite graph is symmetric with respect to 0, a result originally obtained by Coulson and Rushbrooke (1940) in the context of theoretical chemistry.

2d *The derivative of* χ For $i = 1, 2, \ldots, n$ let Γ_i denote the induced subgraph $\langle V\Gamma \setminus v_i \rangle$. Then

$$\chi'(\Gamma; \lambda) = \sum_{i=1}^{n} \chi(\Gamma_i; \lambda).$$

2e *The eigenvalue* 0   Suppose that a graph has two vertices v_i and v_j such that the set of vertices adjacent to v_i is the same as the set of vertices adjacent to v_j. Then the vector $\mathbf{x}$ whose only non-zero components are $x_i = 1$ and $x_j = -1$ is an eigenvector of the adjacency matrix, with eigenvalue 0. If Γ has a set of r vertices, all of which have the same set of neighbours, then the multiplicity of 0 is at least $r - 1$. (An alternative argument uses the observation that there are r equal columns of $\mathbf{A}$, and so its rank is at most $n - r + 1$.)

2f *Cospectral graphs*   Two non-isomorphic graphs are said to be *cospectral* if they have the same eigenvalues with the same multiplicities. The first example of this phenomenon was given by Collatz and Sinogowitz (1957), and many examples are given in [CvDS, pp. 156–161]. Two connected graphs with 6 vertices, both having the characteristic polynomial $\lambda^6 - 7\lambda^4 - 4\lambda^3 + 7\lambda^2 + 4\lambda - 1$, are shown in Figure 1.

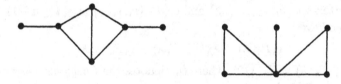

Figure 1: two cospectral graphs.

2g *The walk-generating matrix*   Let $g_{ij}(r)$ denote the number of walks of length r in Γ from v_i to v_j. If we write $\mathbf{G}(z)$ for the matrix

$$(\mathbf{G}(z))_{ij} = \sum_{r=1}^{\infty} g_{ij}(r)\, z^r,$$

then $\mathbf{G}(z) = (\mathbf{I} - z\mathbf{A})^{-1}$, where $\mathbf{A}$ is the adjacency matrix of Γ. This may be regarded as a matrix over the ring of formal power series in z, or as a real matrix defined whenever $z \notin \operatorname{Spec}\Gamma$. From the formula for the inverse matrix and **2e**, we obtain

$$\mathbf{G}(z)_{ii} = \frac{\chi(\Gamma_i; z^{-1})}{z\chi(\Gamma; z^{-1})}, \quad \operatorname{tr}\mathbf{G}(z) = \frac{\chi'(\Gamma; z^{-1})}{z\chi(\Gamma; z^{-1})}.$$

2h *Closed walks and sums of powers of eigenvalues*   A *closed* walk is one whose initial and final vertices coincide. By Lemma 2.5 the total number of closed walks of length l is equal to $\operatorname{tr}\mathbf{A}^l$. Since the trace of a matrix is the sum of its eigenvalues, an alternative expression is $\sum \lambda_i^l$. In particular, the sum of the eigenvalues is zero, the sum of the squares is twice the number of edges, and the sum of the cubes is six times the number of triangles.

2i *An upper bound for the largest eigenvalue*   Suppose that the eigenvalues of Γ are $\lambda_0 \geq \lambda_1 \geq \ldots \geq \lambda_{n-1}$, where Γ has n vertices and m edges. From **2h** we obtain $\sum \lambda_i = 0$ and $\sum \lambda_i^2 = 2m$. It follows that

$$\lambda_0 \leq \left(\frac{2m(n-1)}{n} \right)^{\frac{1}{2}}.$$

Another bound of the same type is $\lambda_0 \leq \sqrt{2m - n + 1}$ (Yuan 1988).

2j *The spectral decomposition* (Godsil and Mohar 1988)   The adjacency matrix has a *spectral decomposition* $\mathbf{A} = \sum \lambda_a \mathbf{E}_a$, where the matrices $\mathbf{E}_a$ are idempotent and mutually orthogonal. It is easy to check that, given a set of mutually orthonormal eigenvectors $\mathbf{x}_a$, we can take

$$\mathbf{E}_a = \mathbf{x}_a \mathbf{x}_a^t, \quad \text{that is} \quad (\mathbf{E}_a)_{ij} = (\mathbf{x}_a)_i (\mathbf{x}_a)_j.$$

It follows that if f is any function for which $f(\mathbf{A})$ is defined, then $f(\mathbf{A}) = \sum f(\lambda_a) \mathbf{E}_a$. For example, the walk-generating matrix $\mathbf{G}(z) = (\mathbf{I} - z\mathbf{A})^{-1}$ is defined whenever $z \notin \mathrm{Spec}\Gamma$, and it can be expressed in the form

$$\mathbf{G}(z) = \sum_{a=0}^{n-1} (1 - \lambda_a z)^{-1} \mathbf{E}_a.$$

This yields the following expression for the individual walk-generating functions:

$$\mathbf{G}(z)_{ij} = \sum_{a=0}^{n-1} (1 - \lambda_a z)^{-1} (\mathbf{x}_a)_i (\mathbf{x}_a)_j.$$

2k *The distance matrices*   For a graph with diameter d the *distance matrices* $\mathbf{A}_h$ ($0 \leq h \leq d$) are defined as follows:

$$(\mathbf{A}_h)_{ij} = \begin{cases} 1, & \text{if } \partial(v_i, v_j) = h; \\ 0, & \text{otherwise.} \end{cases}$$

It follows that

$$\mathbf{A}_0 = \mathbf{I}, \quad \mathbf{A}_1 = \mathbf{A}, \quad \mathbf{A}_0 + \mathbf{A}_1 + \mathbf{A}_2 + \ldots + \mathbf{A}_d = \mathbf{J},$$

where $\mathbf{J}$ is the matrix in which each entry is 1. The distance matrix $\mathbf{A}_h$ can be expressed as a polynomial of degree h in $\mathbf{A}$, for each h in $\{0, 1, \ldots, d\}$, if and only if the graph is *distance-regular* (see Chapter 20). For such a graph the adjacency algebra has the minimum possible dimension $d + 1$.

3

Regular graphs and line graphs

In this chapter we discuss graphs which possess some kinds of combinatorial regularity, and whose spectra, in consequence, have distinctive features. A graph is said to be *regular* of degree k (or *k-regular*) if each of its vertices has degree k. This is the most obvious kind of combinatorial regularity, and it has interesting consequences for the eigenvalues.

Proposition 3.1 *Let Γ be a regular graph of degree k. Then:*
  (1) k *is an eigenvalue of* Γ;
  (2) *if* Γ *is connected, then the multiplicity of k is* 1;
  (3) *for any eigenvalue λ of Γ, we have* $|\lambda| \leq k$.

Proof (1) Let $\mathbf{u} = [1, 1, \ldots, 1]^t$; then if $\mathbf{A}$ is the adjacency matrix of Γ we have $\mathbf{Au} = k\mathbf{u}$, since there are k 1's in each row. Thus k is an eigenvalue of Γ.

  (2) Let $\mathbf{x} = [x_1, x_2, \ldots, x_n]^t$ denote any non-zero vector for which $\mathbf{Ax} = k\mathbf{x}$, and suppose that x_j is an entry of $\mathbf{x}$ with the largest absolute value. Since $(\mathbf{Ax})_j = kx_j$, we have

$$\Sigma' x_i = kx_j$$

where Σ' denotes summation over those k vertices v_i which are adjacent to v_j. By the maximal property of x_j, it follows that $x_i = x_j$ for all these vertices. If Γ is connected we may proceed successively in this way, eventually showing that all entries of $\mathbf{x}$ are equal. Thus $\mathbf{x}$ is a multiple of $\mathbf{u}$, and the space of eigenvectors associated with the eigenvalue k has dimension 1.

  (3) Suppose that $\mathbf{Ay} = \lambda\mathbf{y}, \mathbf{y} \neq \mathbf{0}$, and let y_j denote an entry of $\mathbf{y}$

which is largest in absolute value. By the same argument as in (2), we have $\Sigma' y_i = \lambda y_j$, and so

$$|\lambda||y_j| = |\Sigma' y_i| \leq \Sigma'|y_i| \leq k|y_j|.$$

Thus $|\lambda| \leq k$, as required. $\qquad\square$

The adjacency algebra of a regular connected graph also has a distinctive property, related to the results of Proposition 3.1. Let $\mathbf{J}$ denote the matrix each of whose entries is $+1$. Then, if $\mathbf{A}$ is the adjacency matrix of a regular graph of degree k, we have $\mathbf{AJ} = \mathbf{JA} = k\mathbf{J}$. This is the point of departure for the following result.

Proposition 3.2 (Hoffman 1963)   *The matrix $\mathbf{J}$ belongs to the adjacency algebra $\mathcal{A}(\Gamma)$ if and only if Γ is a regular connected graph.*

Proof   Suppose $\mathbf{J}$ is in $\mathcal{A}(\Gamma)$. By the definition of $\mathcal{A}(\Gamma)$, $\mathbf{J}$ is a polynomial in $\mathbf{A}$; consequently $\mathbf{AJ} = \mathbf{JA}$. Now, if $k^{(i)}$ denotes the degree of the vertex v_i, then $(\mathbf{AJ})_{ij} = k^{(i)}$ and $(\mathbf{AJ})_{ij} = k^{(j)}$, so that all the degrees are equal and Γ is regular. Further, if Γ were disconnected we could find two vertices with no walks joining them, so that the corresponding entry of $\mathbf{A}^l$ would be zero for all $l \geq 0$. Then every polynomial in $\mathbf{A}$ would have a zero entry, contradicting the fact that $\mathbf{J} \in \mathcal{A}(\Gamma)$. Thus Γ is connected.

Conversely, suppose that Γ is connected and regular of degree k. Then, by part (1) of Proposition 3.1, k is an eigenvalue of Γ, and so the minimum polynomial of $\mathbf{A}$ is of the form $p(\lambda) = (\lambda - k)q(\lambda)$. Since $p(\mathbf{A}) = \mathbf{0}$, we obtain $\mathbf{A}q(\mathbf{A}) = kq(\mathbf{A})$; that is, each column of $q(\mathbf{A})$ is an eigenvector of $\mathbf{A}$ corresponding to the eigenvalue k. By part (2) of Proposition 3.1, it follows that each column of $q(\mathbf{A})$ is a multiple of $\mathbf{u}$, and since $q(\mathbf{A})$ is a symmetric matrix, it is a multiple of $\mathbf{J}$. Thus $\mathbf{J}$ is a polynomial in $\mathbf{A}$. $\qquad\square$

Corollary 3.3   *Let Γ be a k-regular connected graph with n vertices, and let the distinct eigenvalues of Γ be $k > \lambda_1 > \ldots > \lambda_{s-1}$. Then if $q(\lambda) = \prod(\lambda - \lambda_i)$, where the product is over the range $1 \leq i \leq s-1$, we have*

$$\mathbf{J} = \left(\frac{n}{q(k)}\right) q(\mathbf{A}).$$

Proof   It follows from the proof of Proposition 3.2 that $q(\mathbf{A}) = \alpha\mathbf{J}$, for some constant α. Now the eigenvalues of $q(\mathbf{A})$ are $q(k)$ and $q(\lambda_i)$ for $1 \leq i \leq s-1$, and all of these except $q(k)$ are zero. The only non-zero eigenvalue of $\alpha\mathbf{J}$ is αn, hence $\alpha = q(k)/n$. $\qquad\square$

For some classes of regular graphs, such as the *strongly regular graphs*
(3c), it is possible to determine the polynomial function f for which
$f(\mathbf{A}) = \mathbf{J}$ by direct means, based on Lemma 2.5. This provides a powerful method for determining the spectra of these graphs. At a more
basic level, there is a special class of regular graphs whose spectra can
be found by means of a well-known technique in matrix theory. As
this class contains several important families of graphs, we shall briefly
review the relevant theory.

An $n \times n$ matrix $\mathbf{S}$ is said to be a *circulant matrix* if its entries satisfy
$s_{ij} = s_{1,j-i+1}$, where the subscripts are reduced modulo n and lie in the
set $\{1, 2, \ldots, n\}$. In other words, row i of $\mathbf{S}$ is obtained from the first
row of $\mathbf{S}$ by a cyclic shift of $i - 1$ steps, and so any circulant matrix is
determined by its first row. Let $\mathbf{W}$ denote the circulant matrix whose
first row is $[0, 1, 0, \ldots, 0]$, and let $\mathbf{S}$ denote a general circulant matrix
whose first row is $[s_1, s_2, \ldots, s_n]$. Then a straightforward calculation
shows that
$$\mathbf{S} = \sum_{j=1}^{n} s_j \mathbf{W}^{j-1}.$$
Since the eigenvalues of $\mathbf{W}$ are $1, \omega, \omega^2, \ldots, \omega^{n-1}$, where $\omega = \exp(2\pi i/n)$,
it follows that the eigenvalues of $\mathbf{S}$ are
$$\lambda_r = \sum_{j=1}^{n} s_j \omega^{(j-1)r}, \quad r = 0, 1, \ldots, n-1.$$

Definition 3.4   A *circulant graph* is a graph Γ whose vertices can be
ordered so that the adjacency matrix $\mathbf{A}(\Gamma)$ is a circulant matrix.

The adjacency matrix is a symmetric matrix with zero entries on the
main diagonal. It follows that, if the first row of the adjacency matrix
of a circulant graph is $[a_1, a_2, \ldots, a_n]$, then $a_1 = 0$ and $a_i = a_{n-i+2}$ for
$i = 2, \ldots, n$.

Proposition 3.5   *Suppose that $[0, a_2, \ldots, a_n]$ is the first row of the
adjacency matrix of a circulant graph Γ. Then the eigenvalues of Γ are*
$$\lambda_r = \sum_{j=2}^{n} a_j \omega^{(j-1)r}, \quad r = 0, 1, \ldots, n-1.$$

Proof   This result follows directly from the expression for the eigenvalues of a circulant matrix.   □

We remark that the n eigenvalues given by the formula of Proposition
3.5 are not necessarily all distinct.

We shall give three examples of this technique. First, the complete graph K_n is a circulant graph: the first row of its adjacency matrix is $[0, 1, 1, \ldots, 1]$. Since

$$1 + \omega^r + \ldots + \omega^{(n-1)r} = 0 \quad \text{for} \quad r \in \{1, 2, \ldots, n-1\},$$

it follows from Proposition 3.5 that the spectrum of K_n is:

$$\text{Spec } K_n = \begin{pmatrix} n-1 & -1 \\ 1 & n-1 \end{pmatrix}.$$

Our second example is the *cycle graph* C_n, whose adjacency matrix is a circulant matrix with first row $[0, 1, 0, \ldots, 0, 1]$. In the notation of Proposition 3.5, the eigenvalues are $\lambda_r = 2\cos(2\pi r/n)$, but these numbers are not all distinct; taking account of coincidences the complete description of the spectrum is:

$$\text{Spec } C_n = \begin{pmatrix} 2 & 2\cos 2\pi/n & \ldots & 2\cos(n-1)\pi/n \\ 1 & 2 & \ldots & 2 \end{pmatrix} \quad (n \text{ odd}),$$

$$\text{Spec } C_n = \begin{pmatrix} 2 & 2\cos 2\pi/n & \ldots & 2\cos(n-2)\pi/n & -2 \\ 1 & 2 & \ldots & 2 & 1 \end{pmatrix} \quad (n \text{ even}).$$

A third family of circulant graphs are the graphs H_s obtained by removing s disjoint edges from K_{2s}. The graph H_s is sometimes known as a *hyperoctahedral graph*, because it is the skeleton of a hyperoctahedron in s dimensions. It is also known as the *cocktail-party graph* $CP(s)$, so called because it is alleged that if there are s married couples at a cocktail party each person talks to everyone except their spouse. It is also a special kind of complete multipartite graph, to be defined in Chapter 6. Clearly, the graph H_s is a circulant graph; we may take the first row of its adjacency matrix to be $[a_1, \ldots, a_{2s}]$, where each entry is 1, except that $a_1 = a_{s+1} = 0$. It follows that the eigenvalues of H_s are

$$\lambda_0 = 2s - 2, \quad \lambda_r = -1 - \omega^{rs} \quad (1 \le r \le 2s - 1),$$

where $\omega^{2s} = 1$ and $\omega \ne 1$. Consequently,

$$\text{Spec } H_s = \begin{pmatrix} 2s-2 & 0 & -2 \\ 1 & s & s-1 \end{pmatrix}.$$

We now turn to another structural property which has implications for the spectrum of a graph. The *line graph* $L(\Gamma)$ of a graph Γ is constructed by taking the edges of Γ as vertices of $L(\Gamma)$, and joining two vertices in $L(\Gamma)$ whenever the corresponding edges in Γ have a common vertex. The spectra of line graphs were investigated extensively by Hoffman (1969) and others. Here we outline the basic results; more recent work is described in the Additional Results at the end of the chapter.

We shall continue to suppose that Γ has n vertices, $v_1, v_2, \ldots, v_n$. We shall need to label the edges of Γ also; that is, $E\Gamma = \{e_1, e_2, \ldots, e_m\}$. For the purposes of this chapter only we define an $n \times m$ matrix $\mathbf{X} = \mathbf{X}(\Gamma)$ as follows:

$$(\mathbf{X})_{ij} = \begin{cases} 1, & \text{if } v_i \text{ and } e_j \text{ are incident;} \\ 0, & \text{otherwise.} \end{cases}$$

Lemma 3.6 *Suppose that Γ and $\mathbf{X}$ are as above. Let $\mathbf{A}$ denote the adjacency matrix of Γ and $\mathbf{A}_L$ the adjacency matrix of $L(\Gamma)$. Then:*
(1) $\mathbf{X}^t\mathbf{X} = \mathbf{A}_L + 2\mathbf{I}_m$;
(2) *if Γ is regular of degree k, then $\mathbf{X}\mathbf{X}^t = \mathbf{A} + k\mathbf{I}_n$.*
The subscripts denote the sizes of the identity matrices.

Proof   (1) We have

$$(\mathbf{X}^t\mathbf{X})_{ij} = \sum_{l=1}^{n} (\mathbf{X})_{li}(\mathbf{X})_{lj},$$

from which it follows that $(\mathbf{X}^t\mathbf{X})_{ij}$ is the number of vertices v_l of Γ which are incident with both the edges e_i and e_j. The required result is now a consequence of the definitions of $L(\Gamma)$ and $\mathbf{A}_L$.

(2) This part is proved by a similar counting argument.   □

Proposition 3.7   *If λ is an eigenvalue of a line graph $L(\Gamma)$, then $\lambda \geq -2$.*

Proof   The matrix $\mathbf{X}^t\mathbf{X}$ is non-negative definite, since we have $\mathbf{z}^t\mathbf{X}^t\mathbf{X}\mathbf{z} = \|\mathbf{X}\mathbf{z}\|^2 \geq 0$, for any vector $\mathbf{z}$. Thus the eigenvalues of $\mathbf{X}^t\mathbf{X}$ are non-negative. But $\mathbf{A}_L = \mathbf{X}^t\mathbf{X} - 2\mathbf{I}_m$, so the eigenvalues of $\mathbf{A}_L$ are not less than -2.   □

The condition that all eigenvalues of a graph be not less than -2 is a restrictive one, but it is not sufficient to characterize line graphs. For example, the hyperoctahedral graphs H_s satisfy this condition, but these graphs are not line graphs. Seidel (1968, see **3g**) gave examples of regular graphs which have least eigenvalue -2 and are neither line graphs nor hyperoctahedral graphs. Subsequently, a characterization of all graphs with least eigenvalue -2 was obtained by Cameron, Goethals, Seidel, and Shult (1976, see **3i**).

When Γ is a regular graph of degree k, its line graph $L(\Gamma)$ is regular of degree $2k - 2$. We can think of this as a connection between the maximum eigenvalues of Γ and $L(\Gamma)$, and in fact the connection extends to all eigenvalues, by virtue of the following result.

Theorem 3.8 (Sachs 1967)   *If Γ is a regular graph of degree k with n vertices and $m = \frac{1}{2}nk$ edges, then*

$$\chi(L(\Gamma); \lambda) = (\lambda + 2)^{m-n}\chi(\Gamma; \lambda + 2 - k).$$

Proof   We shall use the notation and results of Lemma 3.6. Define two partitioned matrices with $n + m$ rows and columns as follows:

$$\mathbf{U} = \begin{bmatrix} \lambda\mathbf{I}_n & -\mathbf{X} \\ 0 & \mathbf{I}_m \end{bmatrix}, \quad \mathbf{V} = \begin{bmatrix} \mathbf{I}_n & \mathbf{X} \\ \mathbf{X}^t & \lambda\mathbf{I}_m \end{bmatrix}.$$

Then we have

$$\mathbf{UV} = \begin{bmatrix} \lambda\mathbf{I}_n - \mathbf{X}\mathbf{X}^t & 0 \\ \mathbf{X}^t & \lambda\mathbf{I}_m \end{bmatrix}, \quad \mathbf{VU} = \begin{bmatrix} \lambda\mathbf{I}_n & 0 \\ \lambda\mathbf{X}^t & \lambda\mathbf{I}_m - \mathbf{X}^t\mathbf{X} \end{bmatrix}.$$

Since $\det(\mathbf{UV}) = \det(\mathbf{VU})$ we deduce that

$$\lambda^m\det(\lambda\mathbf{I}_n - \mathbf{X}\mathbf{X}^t) = \lambda^n\det(\lambda\mathbf{I}_m - \mathbf{X}^t\mathbf{X}).$$

Thus we may argue as follows:

$$\begin{aligned}
\chi(L(\Gamma); \lambda) &= \det(\lambda\mathbf{I}_m - \mathbf{A}_L) \\
&= \det((\lambda + 2)\mathbf{I}_m - \mathbf{X}^t\mathbf{X}) \\
&= (\lambda + 2)^{m-n}\det((\lambda + 2)\mathbf{I}_n - \mathbf{X}\mathbf{X}^t) \\
&= (\lambda + 2)^{m-n}\det((\lambda + 2 - k)\mathbf{I}_n - \mathbf{A}) \\
&= (\lambda + 2)^{m-n}\chi(\Gamma; \lambda + 2 - k).
\end{aligned}$$

$\square$

It follows from Theorem 3.8 that, if the spectrum of Γ is

$$\text{Spec } \Gamma = \begin{pmatrix} k & \lambda_1 & \cdots & \lambda_{s-1} \\ 1 & m_1 & \cdots & m_{s-1} \end{pmatrix},$$

then the spectrum of $L(\Gamma)$ is

$$\text{Spec } L(\Gamma) = \begin{pmatrix} 2k - 2 & k - 2 + \lambda_1 & \cdots & k - 2 + \lambda_{s-1} & -2 \\ 1 & m_1 & \cdots & m_{s-1} & m - n \end{pmatrix}.$$

For example, the line graph $L(K_t)$ is sometimes called the *triangle graph* and denoted by Δ_t. Its vertices correspond to the $\frac{1}{2}t(t-1)$ pairs of numbers from the set $\{1, 2, \ldots, t\}$, two vertices being adjacent whenever the corresponding pairs have just one common member. From the known spectrum of K_t, and Theorem 3.8, we have

$$\text{Spec } \Delta_t = \begin{pmatrix} 2t - 4 & t - 4 & -2 \\ 1 & t - 1 & \frac{1}{2}t(t - 3) \end{pmatrix}.$$

Additional results

3a *The complement of a regular graph*   Let Γ be a graph with n vertices and let Γ^c denote its complement, that is, the graph with the same vertex-set whose edge-set is complementary to that of Γ. Let $\mathbf{A}^c$ denote the adjacency matrix of Γ^c. Then $\mathbf{A} + \mathbf{A}^c = \mathbf{J} - \mathbf{I}$. It was proved by Sachs (1962) that if Γ is connected and regular of degree k, then

$$(\lambda + k + 1)\chi(\Gamma^c; \lambda) = (-1)^n(\lambda - n + k + 1)\chi(\Gamma; -\lambda - 1).$$

3b *The Petersen graph*   The complement of the line graph of K_5 is known as the Petersen graph. It occurs in many contexts throughout graph theory. We shall denote it by the symbol O_3, as it is the case $k = 3$ of the family $\{O_k\}$ of *odd graphs*, to be defined later (**8f**). We have

$$\mathrm{Spec}\ O_3 = \begin{pmatrix} 3 & 1 & -2 \\ 1 & 5 & 4 \end{pmatrix}.$$

In particular, the least eigenvalue is -2, although O_3 is neither a line graph nor a hyperoctahedral graph.

3c *Strongly regular graphs*   A k-regular graph is said to be *strongly regular*, with parameters (k, a, c), if the following conditions hold. Each pair of adjacent vertices has the same number $a \geq 0$ of common neighbours, and each pair of non-adjacent vertices has the same number $c \geq 1$ of common neighbours. It follows from Lemma 2.5 that the adjacency matrix of such a graph satisfies

$$\mathbf{A}^2 + (c - a)\mathbf{A} + (c - k)\mathbf{I} = c\mathbf{J}.$$

In other words, the polynomial function f whose existence is guaranteed by Proposition 3.2 is $f(x) = (1/c)(x^2 + (c - a)x + (c - k))$.

3d *The spectrum of a strongly regular graph*   Since the eigenvalues of the $n \times n$ matrix $\mathbf{J}$ are n (with multiplicity 1) and 0 (with multiplicity $n - 1$), it follows from **3c** that the eigenvalues of a strongly regular graph are k (with multiplicity 1) and the two roots λ_1, λ_2 of the quadratic equation $f(\lambda) = 0$ (with total multiplicity $n - 1$). The multiplicities $m_1 = m(\lambda_1)$ and $m_2 = m(\lambda_2)$ can be determined from the equations

$$m_1 + m_2 = n - 1, \quad k + m_1\lambda_1 + m_2\lambda_2 = 0,$$

the second of which follows from **2h**. For example, the Petersen graph (**3b**) is strongly regular with parameters $(3, 0, 1)$, and this gives an alternative method of determining its spectrum.

3e *The Möbius ladders*   The Möbius ladder M_h is a regular graph of degree 3 with $2h$ vertices ($h \geq 3$). It is constructed from the cycle graph

C_{2h} by adding new edges joining each pair of 'opposite' vertices, and so it is a circulant graph. The eigenvalues are the numbers

$$\lambda_j = 2\cos(\pi j/h) + (-1)^j \quad (0 \le j \le 2h - 1).$$

3f *Graphs characterized by their spectra*   Although there are many examples of cospectral graphs, there are also cases where there is a unique graph with a given spectrum. We give two instances.

(*a*) The spectrum of the triangle graph $\Delta_t = L(K_t)$ is given above. If Γ is a graph for which $\operatorname{Spec}\Gamma = \operatorname{Spec}\Delta_t$, and $t \ne 8$, then $\Gamma = \Delta_t$. In the case $t = 8$ there are three exceptional graphs, not isomorphic with Δ_8, but having the same spectrum as Δ_8 (Chang 1959, Hoffman 1960).

(*b*) The *complete bipartite graph* $K_{a,a}$ is constructed by taking two sets of a vertices and joining every vertex in the first set to every vertex in the second. If Γ is a graph for which $\operatorname{Spec}\Gamma = \operatorname{Spec}L(K_{a,a})$, and $a \ne 4$, then $\Gamma = L(K_{a,a})$. In the case $a = 4$ there is one exceptional graph; this graph is depicted in Figure 2 (Shrikhande 1959).

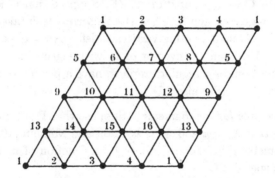

Figure 2: Shrikhande's exceptional graph.

3g *Regular graphs with least eigenvalue* -2 The following graphs having least eigenvalue -2 were noted by Seidel (1968). They are neither line graphs nor hyperoctahedral graphs:

(a) the Petersen graph;
(b) a 5-regular graph with 16 vertices;
(c) a 16-regular graph with 27 vertices (see p. 57);
(d) the exceptional graphs mentioned in **3f**.

3h *Generalized line graphs*   The cocktail party graph $CP(s)$ is defined on page 17. For any graph Γ with vertices $v_1, v_2, \ldots, v_n$, and any non-negative integers $a_1, a_2, \ldots, a_n$, we construct the *generalized line graph* $L(\Gamma; a_1, a_2, \ldots, a_n)$ as follows. The vertex-set is the union of the vertex-sets of $L(\Gamma), CP(a_1), CP(a_2), \ldots, CP(a_n)$, and the edge-set is the union of the edge-sets, together with edges joining all vertices of $CP(a_i)$ to every vertex of $L(\Gamma)$ corresponding to an edge of Γ containing v_i, for $1 \leq i \leq n$. A generalized line graph constructed in this way has least eigenvalue -2.

3i *All graphs with least eigenvalue* -2 If Γ is a graph with least eigenvalue not less than -2, then $\mathbf{A} + 2\mathbf{I}$ is non-negative definite, and so $\mathbf{A} + 2\mathbf{I} = \mathbf{MM}^t$ for some matrix $\mathbf{M}$. By establishing a correspondence between the rows of $\mathbf{M}$ and sets of vectors known as *root systems*, Cameron, Goethals, Seidel, and Shult (1976) showed that all graphs with least eigenvalue not less than -2 fall into three classes: (a) the line graphs of bipartite graphs; (b) the generalized line graphs described in **3h**; (c) a finite class of graphs arising from the root systems E_6, E_7, E_8.

3j *Perfect codes in regular graphs*   For any vertex v of a graph Γ define the *e-neighbourhood* of v to be $N_e(v) = \{u \in V\Gamma \mid \partial(u, v) \leq e\}$. A *perfect e-code* in Γ is a set $C \subseteq V\Gamma$ such that the e-neighbourhoods $N_e(c)$ with $c \in C$ form a partition of $V\Gamma$. Suppose that C is a perfect 1-code in a k-regular graph Γ. Then the vector $\mathbf{c}$ which takes the value 1 on vertices in C and 0 on other vertices satisfies $\mathbf{Ac} = \mathbf{u} - \mathbf{c}$. It follows that $\mathbf{u} - (k+1)\mathbf{c}$ is an eigenvector of Γ, with eigenvalue -1. Thus a necessary condition for a regular graph to have a perfect 1-code is that -1 is an eigenvalue. (See also **21j**.)

3k *Spectral bounds for the diameter*   Suppose that Γ is connected and k-regular, so that $\lambda_0 = k$ and $\lambda_1 < k$. Alon and Milman (1985) proved that the diameter d is bounded above by a function of n, k, and the 'gap' $k - \lambda_1$; specifically

$$d \leq 2 \left\lceil \left(\frac{2k}{k - \lambda_1} \right)^{\frac{1}{2}} \log_2 n \right\rceil.$$

Mohar (1991) improved this to

$$d \leq 2 \left\lceil \left(\frac{2k - \lambda_1}{4(k - \lambda_1)} \right) \ln(n - 1) \right\rceil.$$

Using the spectral decomposition of $\mathbf{A}^r$ (see **2j**), Chung (1989) obtained a bound involving the second largest eigenvalue *in absolute value* $\lambda_* = \max(\lambda_1, -\lambda_{n-1})$. She showed that if $(k/\lambda_*)^r > n - 1$ then $(\mathbf{A}^r)_{ij} > 0$ for all i, j. It follows that

$$d \leq \left\lceil \frac{\ln(n - 1)}{\ln(k/\lambda_*)} \right\rceil.$$

4

Cycles and cuts

Let $\mathbb{C}$ denote the field of complex numbers, and let X be any finite set. Then the set of all functions from X to $\mathbb{C}$ has the structure of a finite-dimensional vector space; if $f : X \to \mathbb{C}$ and $g : X \to \mathbb{C}$, then the vector space operations are defined by the rules

$$(f + g)(x) = f(x) + g(x), \quad (\alpha f)(x) = \alpha f(x) \quad (x \in X, \alpha \in \mathbb{C}).$$

The dimension of this vector space is equal to the number of members of X.

Definition 4.1 The *vertex-space* $C_0(\Gamma)$ of a graph is the vector space of all functions from $V\Gamma$ to $\mathbb{C}$. The *edge-space* $C_1(\Gamma)$ of Γ is the vector space of all functions from $E\Gamma$ to $\mathbb{C}$.

Taking $V\Gamma = \{v_1, v_2, \ldots, v_n\}$ and $E\Gamma = \{e_1, e_2, \ldots, e_m\}$, it follows that $C_0(\Gamma)$ is a vector space of dimension n and $C_1(\Gamma)$ is a vector space of dimension m. Any function $\eta : V\Gamma \to \mathbb{C}$ can be represented by a column vector

$$\mathbf{y} = [y_1, y_2, \ldots, y_n]^t,$$

where $y_i = \eta(v_i)$ $(1 \leq i \leq n)$. This representation corresponds to choosing as a basis for $C_0(\Gamma)$ the set of functions $\{\omega_1, \omega_2, \ldots, \omega_n\}$, defined by

$$\omega_i(v_j) = \begin{cases} 1, & \text{if } i = j; \\ 0, & \text{otherwise.} \end{cases}$$

In a similar way, we may choose the basis $\{\epsilon_1, \epsilon_2, \ldots, \epsilon_m\}$ for $C_1(\Gamma)$

defined by

$$\epsilon_i(e_j) = \begin{cases} 1, & \text{if } i = j; \\ 0, & \text{otherwise}; \end{cases}$$

and hence represent a function $\xi : E\Gamma \to \mathbb{C}$ by a column vector $\mathbf{x} = [x_1, x_2, \ldots, x_m]^t$ such that $x_i = \xi(e_i)$ $(1 \le i \le m)$. We shall refer to the bases $\{\omega_1, \omega_2, \ldots, \omega_n\}$ and $\{\epsilon_1, \epsilon_2, \ldots, \epsilon_m\}$ as the *standard bases* for $C_0(\Gamma)$ and $C_1(\Gamma)$.

We now introduce a useful device. For each edge $e_\alpha = \{v_\sigma, v_\tau\}$ of Γ, we shall choose one of v_σ, v_τ to be the *positive end* of e_α, and the other one to be the *negative end*. We refer to this procedure by saying that Γ has been given an *orientation*. Although this device is employed in the proofs of several results, the results themselves are independent of it.

Definition 4.2   The *incidence matrix* $\mathbf{D}$ of Γ, with respect to a given orientation of Γ, is the $n \times m$ matrix (d_{ij}) whose entries are

$$d_{ij} = \begin{cases} +1, & \text{if } v_i \text{ is the positive end of } e_j; \\ -1, & \text{if } v_i \text{ is the negative end of } e_j; \\ 0, & \text{otherwise}. \end{cases}$$

The rows of the incidence matrix correspond to the vertices of Γ, and its columns correspond to the edges of Γ; each column contains just two non-zero entries, $+1$ and -1, representing the positive and negative ends of the corresponding edge.

We remark that $\mathbf{D}$ is the representation, with respect to the standard bases, of a linear mapping from $C_1(\Gamma)$ to $C_0(\Gamma)$. This mapping will be called the *incidence mapping*, and be denoted by D. For each $\xi : E\Gamma \to \mathbb{C}$ the function $D\xi : V\Gamma \to \mathbb{C}$ is defined by

$$D\xi(v_i) = \sum_{j=1}^{m} d_{ij}\xi(e_j) \quad (1 \le i \le n).$$

For the rest of this chapter we shall let c denote the number of connected components of Γ.

Proposition 4.3   *The incidence matrix $\mathbf{D}$ of Γ has rank $n - c$.*

Proof   The incidence matrix can be written in the partitioned form

$$\begin{bmatrix} \mathbf{D}^{(1)} & \mathbf{0} & \cdots & \mathbf{0} \\ \mathbf{0} & \mathbf{D}^{(2)} & \cdots & \mathbf{0} \\ \vdots & \vdots & & \vdots \\ \mathbf{0} & \mathbf{0} & \cdots & \mathbf{D}^{(c)} \end{bmatrix},$$

by a suitable labelling of the vertices and edges of Γ, where the matrix $\mathbf{D}^{(i)}$ $(1 \le i \le c)$ is the incidence matrix of a component $\Gamma^{(i)}$ of Γ. We shall show that the rank of $\mathbf{D}^{(i)}$ is $n_i - 1$, where $n_i = |V\Gamma^{(i)}|$, from which the required result follows by addition.

Let $\mathbf{d}_j$ denote the row of $\mathbf{D}^{(i)}$ corresponding to the vertex v_j of $\Gamma^{(i)}$. Since there is just one $+1$ and just one -1 in each column of $\mathbf{D}^{(i)}$, it follows that the sum of the rows of $\mathbf{D}^{(i)}$ is the zero row vector, and that the rank of $\mathbf{D}^{(i)}$ is at most $n_i - 1$. Suppose we have a linear relation $\sum \alpha_j \mathbf{d}_j = \mathbf{0}$, where the summation is over all rows of $\mathbf{D}^{(i)}$, and not all the coefficients α_j are zero. Choose a row $\mathbf{d}_k$ for which $\alpha_k \neq 0$; this row has non-zero entries in those columns corresponding to the edges incident with v_k. For each such column, there is just one other row $\mathbf{d}_l$ with a non-zero entry in that column, and in order that the given linear relation should hold, we must have $\alpha_l = \alpha_k$. Thus, if $\alpha_k \neq 0$, then $\alpha_l = \alpha_k$ for all vertices v_l adjacent to v_k. Since $\Gamma^{(i)}$ is connected, it follows that all coefficients α_j are equal, and so the given linear relation is just a multiple of $\sum \mathbf{d}_j = \mathbf{0}$. Consequently, the rank of $\mathbf{D}^{(i)}$ is $n_i - 1$.

$\square$

The following definition applies to a general graph Γ with n vertices, m edges, and c components, although for the time being we shall continue to deal with strict graphs, rather than general graphs.

Definition 4.4   The *rank* of Γ and the *co-rank* of Γ are, respectively,

$$r(\Gamma) = n - c; \quad s(\Gamma) = m - n + c.$$

We now investigate the kernel of the incidence mapping D, and its relationship with graph-theoretical properties of Γ. Let Q be a set of edges such that the subgraph $\langle Q \rangle$ is a cycle graph. We say that Q is a *cycle in* Γ; the two possible cyclic orderings of the vertices of $\langle Q \rangle$ induce two possible *cycle-orientations* of the edges Q. Let us choose one of these cycle-orientations, and define a function ξ_Q in $C_1(\Gamma)$ as follows. We put $\xi_Q(e) = +1$ if e belongs to Q and its cycle-orientation coincides with its orientation in Γ, $\xi_Q(e) = -1$ if e belongs to Q and its cycle-orientation is the reverse of its orientation in Γ, while if e is not in Q we put $\xi_Q(e) = 0$.

Theorem 4.5   *The kernel of the incidence mapping D of Γ is a vector space whose dimension is equal to the co-rank of Γ. If Q is a cycle in Γ, then ξ_Q belongs to the kernel of D.*

Proof   Since the rank of D is $n - c$, and the dimension of $C_1(\Gamma)$ is m, it follows that the kernel of D has dimension $m - n + c = s(\Gamma)$. With respect to the standard bases for $C_1(\Gamma)$ and $C_0(\Gamma)$, we may take D to be the incidence matrix, and ξ_Q to be represented by a column vector $\mathbf{x}_Q$. Now $(\mathbf{D}\mathbf{x}_Q)_i$ is the inner product of the row $\mathbf{d}_i$ of $\mathbf{D}$ and the vector $\mathbf{x}_Q$. If v_i is not incident with some edges of Q, then this inner product is 0; if v_i is incident with some edges of Q, then it is incident with precisely two edges, and the choice of signs in the definition of ξ_Q implies that the inner product is again 0. Thus $\mathbf{D}\mathbf{x}_Q = \mathbf{0}$, and ξ_Q belongs to the kernel of D. □

If ρ and σ are two elements of the edge-space of Γ (that is, functions from $E\Gamma$ to $\mathbb{C}$), then we may define their inner product

$$(\rho, \sigma) = \sum_{e \in E\Gamma} \rho(e)\overline{\sigma(e)},$$

where the overline indicates the complex conjugate. When ρ and σ are represented by coordinate vectors, with respect to the standard basis of $C_1(\Gamma)$, this inner product corresponds to the usual inner product of vectors in the complex vector space $\mathbb{C}^m$. (In practice we use only functions with real values, so the conjugation is irrelevant.)

Definition 4.6 The *cycle-subspace* of Γ is the kernel of the incidence mapping of Γ. The *cut-subspace* of Γ is the orthogonal complement of the cycle-subspace in $C_1(\Gamma)$, with respect to the inner product defined above.

The first part of this definition is justified by the result of Theorem 4.5, which says that vectors representing cycles belong to the cycle-subspace: indeed, in the next chapter we shall show how to construct a basis for the cycle-subspace consisting entirely of cycles. We now proceed to justify the second part of the definition.

Let $V\Gamma = V_1 \cup V_2$ be a partition of $V\Gamma$ into non-empty disjoint subsets. If the set H of edges of Γ which have one vertex in V_1 and one vertex in V_2 is non-empty, then we say that H is a *cut* in Γ. We may choose one of the two possible *cut-orientations* for H, by specifying that one of V_1, V_2 contains the positive ends of all edges in H, while the other contains the negative ends. We now define a function ξ_H in $C_1(\Gamma)$ by putting $\xi_H(e) = +1$ if e belongs to H and its cut-orientation coincides with its orientation in Γ, $\xi_H(e) = -1$ if e belongs to H and its cut-orientation is the reverse of its orientation in Γ, and $\xi_H(e) = 0$ if e is not in H.

Proposition 4.7 *The cut-subspace of Γ is a vector space whose dimension is equal to the rank of Γ. If H is a cut in Γ, then ξ_H belongs to the cut-subspace.*

Proof Since the dimension of the cycle-subspace is $m - n + c$, its orthogonal complement, the cut-subspace, has dimension $n - c = r(\Gamma)$.

If H is a cut in Γ, we have $V\Gamma = V_1 \cup V_2$, where V_1 and V_2 are disjoint and non-empty, and H consists precisely of those edges which have one vertex in V_1 and one vertex in V_2. Thus, if $\mathbf{x}_H$ is the column vector representing ξ_H, we have

$$\mathbf{x}_H^t = \pm \frac{1}{2} \left[\sum_{v_i \in V_1} \mathbf{d}_i - \sum_{v_i \in V_2} \mathbf{d}_i \right],$$

where $\mathbf{d}_i$ is the row of the incidence matrix corresponding to v_i. The sign on the right-hand side of this equation depends only on which of the two possible cut-orientations has been chosen for H. Now if $\mathbf{Dz} = \mathbf{0}$, then $\mathbf{d}_i\mathbf{z} = 0$ for each $v_i \in V$, and we deduce that $\mathbf{x}_H^t\mathbf{z} = 0$. In other words, ξ_H belongs to the orthogonal complement of the cycle-subspace, and, by definition, this is the cut-subspace. □

The proof of Proposition 4.7 indicates one way of choosing a basis $\{\xi_1, \xi_2, \ldots, \xi_{n-c}\}$ for the cut-subspace of Γ. The set of edges incident with a vertex v_j of Γ forms a cut whose representative vector is $\mathbf{d}_j^t$. If, for each component $\Gamma^{(i)}$ $(1 \le i \le c)$ of Γ, we delete one row of $\mathbf{D}$ corresponding to a vertex in $\Gamma^{(i)}$, then the remaining $n - c$ rows are linearly independent. Furthermore, the transpose of any vector $\mathbf{x}_H$, representing a cut H, can be expressed as a linear combination of these $n - c$ rows, by using the equation displayed in the proof of Proposition 4.7 and the fact that the sum of rows corresponding to each component is $\mathbf{0}$.

This basis has the desirable property that each member represents an actual cut, rather than a 'linear combination' of cuts. It is, however, rather clumsy to work with, and in the next chapter we shall investigate a more elegant procedure which has the added advantage that it provides a basis for the cycle-subspace as well.

We end this chapter by proving a simple relationship between the *Laplacian matrix* $\mathbf{Q} = \mathbf{DD}^t$ and the adjacency matrix of Γ.

Proposition 4.8 *Let $\mathbf{D}$ be the incidence matrix (with respect to some orientation) of a graph Γ, and let $\mathbf{A}$ be the adjacency matrix of Γ. Then the Laplacian matrix $\mathbf{Q}$ satisifies*

$$\mathbf{Q} = \mathbf{DD}^t = \boldsymbol{\Delta} - \mathbf{A},$$

where $\boldsymbol{\Delta}$ is the diagonal matrix whose ith diagonal entry is the degree of the vertex v_i $(1 \le i \le n)$. Consequently, $\mathbf{Q}$ is independent of the orientation given to Γ.

Proof $(\mathbf{DD}^t)_{ij}$ is the inner product of the rows $\mathbf{d}_i$ and $\mathbf{d}_j$ of $\mathbf{D}$. If $i \ne j$, then these rows have a non-zero entry in the same column if and only if there is an edge joining v_i and v_j. In this case, the two non-zero entries are $+1$ and -1, so that $(\mathbf{DD}^t)_{ij} = -1$. Similarly, $(\mathbf{DD}^t)_{ii}$ is the inner product of $\mathbf{d}_i$ with itself, and, since the number of entries ± 1 in $\mathbf{d}_i$ is equal to the degree of v_i, the result follows. □

Additional Results

4a *The coboundary mapping*   The linear mapping from $C_0(\Gamma)$ to $C_1(\Gamma)$ defined (with respect to the standard bases) by $\mathbf{x} \mapsto \mathbf{D}^t\mathbf{x}$ is sometimes called the *coboundary mapping* for Γ. The kernel of the coboundary mapping is a vector space of dimension c, and the image of the coboundary mapping is the cut-subspace of Γ.

4b *The isoperimetric number*   For any set $X \subseteq V\Gamma$ the cut defined by the partition of $V\Gamma$ into X and its complement is denoted by δX. The *isoperimetric number* of Γ is defined to be

$$i(\Gamma) = \min_{|X| \le |V\Gamma|/2} \frac{|\delta X|}{|X|}.$$

For example, it is easy to check that $i(K_n) = \lceil n/2 \rceil$, $i(O_3) = 1$.

4c *Small cycles*   The *girth* of a graph is the number g of edges in a smallest cycle. For example, $g(K_n) = 3$ $(n \ge 3)$, $g(K_{a,a}) = 4$ $(a \ge 2)$, and $g(O_3) = 5$. If Γ has girth $g \ge 2r + 1$ then for each pair of vertices v and w such that $\partial(v,w) = q \le r$ there is a unique walk of length q from v to w. In the k-regular case this leads to the following relations between the adjacency matrix and the distance matrices $\mathbf{A}_q$ $(2 \le q \le r)$ defined in **2k**:

$$\mathbf{A}_2 = \mathbf{A}^2 - k\mathbf{I}, \quad \mathbf{A}_q = \mathbf{A}\mathbf{A}_{q-1} - (k-1)\mathbf{A}_{q-2} \ (3 \le q \le r).$$

It follows that a distance matrix $\mathbf{A}_q$ with $q \le r$ is expressible as a polynomial in $\mathbf{A}$. Explicitly, $\mathbf{A}_q = f_q(\mathbf{A})$, where

$$f_0(x) = 1, \quad f_1(x) = x, \quad f_2(x) = x^2 - k,$$

$$f_q(x) = xf_{q-1}(x) - (k-1)f_{q-2}(x) \quad (q \ge 3).$$

4d *Girth and excess*   It is an elementary exercise (see Chapter 23) to show that the number of vertices in a k-regular graph with girth $g = 2r + 1$ is at least

$$n_0(k,g) = 1 + k + k(k-1) + k(k-1)^2 + \ldots + k(k-1)^r.$$

The Petersen graph O_3 achieves the lower bound n_0 for the case $k = 3$ and $g = 5$, but in the general case graphs which achieve the lower bound are rare (Chapter 23). For any k-regular graph Γ with girth g we define the *excess* to be the amount e by which the lower bound is exceeded, that is, $e = n - n_0(k,g)$, where n is the number of vertices in Γ. Using the equations given in **4c**, Biggs (1980) established a lower bound for e in terms of the eigenvalues of Γ. Define the polynomials g_i by

$$g_i(x) = f_0(x) + f_1(x) + \ldots + f_i(x),$$

where the polynomials f_i are defined above. Then for any k-regular graph with girth $g = 2r + 1$ the excess e satisfies

$$e \geq |g_r(\lambda)|, \quad (\lambda \in \text{Spec } \Gamma, \ \lambda \neq k).$$

4e *The Laplacian spectrum*   Let $\mu_0 \leq \mu_1 \leq \ldots \leq \mu_{n-1}$ be the eigenvalues of the Laplacian matrix $\mathbf{Q}$. Then:
    (a) $\mu_0 = 0$, with eigenvector $[1, 1, \ldots, 1]$;
    (b) if Γ is connected, $\mu_1 > 0$;
    (c) if Γ is regular of degree k, then $\mu_i = k - \lambda_i$, where the λ_i are the (ordinary) eigenvalues of Γ, in weakly decreasing order.

4f *Planar graphs and duality*   A *planar* graph is one which can be drawn in the plane in the usual way, without extraneous crossings of the edges. The *dual* of a graph so drawn is the graph whose vertices are the resulting regions of the plane, two being adjacent when they have a common edge. Let Γ be a connected planar graph, and Γ^* a dual of Γ. If Γ is given an orientation and $\mathbf{D}$ is the incidence matrix of Γ, then Γ^* can be given an orientation so that its incidence matrix $\mathbf{D}^*$ satisfies:
    (a) rank $(\mathbf{D})$ + rank $(\mathbf{D}^*)$ = $|E\Gamma|$;
    (b) $\mathbf{D}^*\mathbf{D}^t = \mathbf{0}$.

4g *The image of the incidence mapping*   Let ω be an element of $C_0(\Gamma)$, where Γ is a connected graph. Then ω is in the image of D if and only if

$$\sum_{v \in V\Gamma} \omega(v) = 0.$$

A more sophisticated way of expressing this result is as follows. Let $S : C_0(\Gamma) \to \mathbb{C}$ denote the linear map defined by $S(\omega) = \sum \omega(v)$; this is known as the *augmentation* map. Then the sequence of linear maps

$$C_1(\Gamma) \xrightarrow{D} C_0(\Gamma) \xrightarrow{S} \mathbb{C} \longrightarrow 0$$

is *exact*. In particular, this means that the image of D is equal to the kernel of S.

4h *Flows*   An element ϕ of the cycle-subspace of Γ is called a *flow* on Γ. The *support* of ϕ, written $S(\phi)$, is the set of edges e for which $\phi(e) \neq 0$; a subset S of $E\Gamma$ is a *minimal support* if $S = S(\phi)$ for some flow ϕ, and the only flow whose support is properly contained in S is the zero flow. We have the following basic facts.
    (a) The set of flows with a given minimal support (together with the zero flow) forms a one-dimensional space.

(b) A minimal support is a cycle.

(c) If ϕ is a flow whose support is minimal, then $|\phi(e)|$ is constant on $S(\phi)$.

4i *Integral flows*   The flow ϕ is *integral* if each $\phi(e)$ is an integer; it is *primitive* if $S(\phi)$ is minimal and each $\phi(e)$ is 0,1 or -1. We say that the flow θ *conforms* to the flow χ if $S(\theta) \subseteq S(\chi)$ and $\theta(e)\chi(e) > 0$ for e in $S(\theta)$. Tutte (1956) showed that:

(a) for a given integral flow ϕ there is a primitive flow which conforms to ϕ;

(b) any integral flow ϕ is the sum of integer multiples of primitive flows, each of which conforms to ϕ.

4j *Modular flows*   Suppose the entries 0,1, -1 of $\mathbf{D}$ are taken to be elements of the ring $\mathbf{Z}_u = \mathbf{Z}/u\mathbf{Z}$ of residue classes of integers modulo u. A *flow mod* u on Γ is a vector $\mathbf{x}$ with components in $\mathbf{Z}_u$ for which $\mathbf{Dx} = \mathbf{0}$, where $\mathbf{0}$ is the zero vector over $\mathbf{Z}_u$. The results in **4i** imply that, if $\mathbf{x}$ is a given flow mod u, then there is an integral flow $\mathbf{y}$, each of whose components y_i satisfies $y_i \in x_i$ and $-u < y_i < u$. Consequently, if Γ has a flow mod u, then it has a flow mod $(u+1)$ (Tutte 1956).

4k *The 5-flow conjecture*   A *nowhere-zero f-flow* ϕ on Γ is a flow mod f for which $S(\phi) = E\Gamma$. Tutte (1954) conjectured that every graph with no isthmus has a nowhere-zero 5-flow. (An *isthmus* is a cut consisting of a single edge.) The following results are known.

(a) Every planar graph with no isthmus has a nowhere-zero 4-flow.

(b) The Petersen graph does not have a nowhere-zero 4-flow.

(c) Every graph with no isthmus has a nowhere-zero 6-flow (Seymour 1981).

5

Spanning trees and associated structures

The problem of finding bases for the cycle-subspace and the cut-subspace is of great practical and theoretical importance. It was originally solved by Kirchhoff (1847) in his studies of electrical networks, and we shall give a brief exposition of that topic at the end of the chapter.

We shall restrict our attention to connected graphs, because the cycle-subspace and the cut-subspace of a disconnected graph are the direct sums of the corresponding spaces for the components. Throughout this chapter, Γ will denote a connected graph with n vertices and m edges, so that $r(\Gamma) = n - 1$ and $s(\Gamma) = m - n + 1$. We shall also assume that Γ has been given an orientation.

A *spanning tree* in Γ is a subgraph which has $n - 1$ edges and contains no cycles. It follows that a spanning tree is connected. We shall use the symbol T to denote both the spanning tree itself and its edge-set. The following simple lemma is a direct consequence of the definition.

Lemma 5.1 *Let T be a spanning tree in a connected graph Γ. Then:*

(1) for each edge g of Γ which is not in T there is a unique cycle in Γ containing g and edges in T only.

(2) for each edge h of Γ which is in T, there is a unique cut in Γ containing h and edges not in T only. $\qquad\square$

We write $\mathrm{cyc}(T, g)$ and $\mathrm{cut}(T, h)$ for the unique cycle and cut whose existence is guaranteed by Lemma 5.1. We give $\mathrm{cyc}(T, g)$ and $\mathrm{cut}(T, h)$ the cycle-orientation and cut-orientation which coincide, on g and h respectively, with the orientation in Γ. Then we have elements $\xi_{(T,g)}$

and $\xi_{(T,h)}$ of the edge-space $C_1(\Gamma)$; these elements are defined (in terms of the given cycle and cut) as in Chapter 4.

Theorem 5.2 *With the same hypothesis as in Lemma 5.1, we have:*

(1) *as g runs through the set $E\Gamma - T$, the $m - n + 1$ elements $\xi_{(T,g)}$ form a basis for the cycle-subspace of Γ.*

(2) *as h runs through the set T, the $n - 1$ elements $\xi_{(T,h)}$ form a basis for the cut-subspace of Γ.*

Proof (1) Since the elements $\xi_{(T,g)}$ correspond to cycles, it follows from Theorem 4.5 that they belong to the cycle-subspace. They form a linearly independent set, because a given edge g in $E\Gamma - T$ belongs to $\text{cyc}(T, g)$ but to no other $\text{cyc}(T, g')$ for $g' \neq g$. Finally, since there are $m - n + 1$ of these elements, and this is the dimension of the cycle-subspace, it follows that we have a basis.

(2) This is proved by arguments analogous to those used in the proof of the first part. $\qquad\square$

We shall now put the foregoing ideas into a form which will show explicitly how cycles and cuts can be derived from the incidence matrix, by means of simple matrix operations. To do this, we shall require some properties of submatrices of the incidence matrix.

Proposition 5.3 (Poincaré 1901) *Any square submatrix of the incidence matrix $\mathbf{D}$ of a graph Γ has determinant equal to 0 or +1 or −1.*

Proof Let $\mathbf{S}$ denote a square submatrix of $\mathbf{D}$. If every column of $\mathbf{S}$ has two non-zero entries, then these entries must be +1 and −1 and so, since each column has sum zero, $\mathbf{S}$ is singular and $\det \mathbf{S} = 0$. Also, if every column of $\mathbf{S}$ has no non-zero entries, then $\det \mathbf{S} = 0$.

The remaining case occurs when a column of $\mathbf{S}$ has precisely one non-zero entry. In this case we can expand $\det \mathbf{S}$ in terms of this column, obtaining $\det \mathbf{S} = \pm \det \mathbf{S}'$, where $\mathbf{S}'$ has one row and column fewer than $\mathbf{S}$. Continuing this process, we eventually arrive at either a zero determinant or a single entry of $\mathbf{D}$, and so the result is proved. $\qquad\square$

Proposition 5.4 *Let U be a subset of $E\Gamma$ with $|U| = n - 1$. Let $\mathbf{D}_U$ denote an $(n-1) \times (n-1)$ submatrix of $\mathbf{D}$, consisting of the intersection of those $n - 1$ columns of $\mathbf{D}$ corresponding to the edges in U and any set of $n - 1$ rows of $\mathbf{D}$. Then $\mathbf{D}_U$ is invertible if and only if the subgraph $\langle U \rangle$ is a spanning tree of Γ.*

Proof Suppose that $\langle U \rangle$ is a spanning tree of Γ. Then the submatrix $\mathbf{D}_U$ consists of $n - 1$ rows of the incidence matrix $\mathbf{D}'$ of U. Since $\langle U \rangle$ is connected, the rank of $\mathbf{D}'$ is $n - 1$, and so $\mathbf{D}_U$ is invertible.

Conversely, suppose that $\mathbf{D}_U$ is invertible. Then the incidence matrix $\mathbf{D}'$ of $\langle U \rangle$ has an invertible $(n-1) \times (n-1)$ submatrix, and consequently the rank of $\mathbf{D}'$ is $(n-1)$. Since $|U| = n-1$, this means that the cycle-subspace of $\langle U \rangle$ has dimension zero, and so $\langle U \rangle$ is a spanning tree of Γ. □

Suppose that $V\Gamma = \{v_1, v_2, \ldots, v_n\}$ and $E\Gamma = \{e_1, e_2, \ldots, e_m\}$, where the labelling has been chosen so that $e_1, e_2, \ldots, e_{n-1}$ are the edges of a given spanning tree T of Γ. The incidence matrix of Γ is then partitioned as follows:

$$ \mathbf{D} = \begin{bmatrix} \mathbf{D}_T & \mathbf{D}_N \\ & \mathbf{d}_n \end{bmatrix}, $$

where $\mathbf{D}_T$ is an $(n-1) \times (n-1)$ square matrix, invertible by Proposition 5.4, and the last row $\mathbf{d}_n$ is linearly dependent on the other rows.

Let $\mathbf{C}$ denote the matrix whose columns are the vectors representing the elements $\xi_{(T, e_j)}$ $(n \leq j \leq m)$ with respect to the standard basis of $C_1(\Gamma)$. Then $\mathbf{C}$ can be written in the partitioned form

$$ \mathbf{C} = \begin{bmatrix} \mathbf{C}_T \\ \mathbf{I}_{m-n+1} \end{bmatrix}. $$

Since every column of $\mathbf{C}$ represents a cycle, and consequently belongs to the kernel of $\mathbf{D}$, we have $\mathbf{DC} = \mathbf{0}$. Thus

$$ \mathbf{C}_T = -\mathbf{D}_T^{-1} \mathbf{D}_N. $$

In a similar fashion, the matrix $\mathbf{K}$ whose columns represent the elements $\xi_{(T, e_j)}$ $(1 \leq j \leq n-1)$ can be written in the form

$$ \mathbf{K} = \begin{bmatrix} \mathbf{I}_{n-1} \\ \mathbf{K}_T \end{bmatrix}. $$

Since each column of $\mathbf{K}$ belongs to the orthogonal complement of the cycle-subspace, we have $\mathbf{CK}^t = \mathbf{0}$; that is, $\mathbf{C}_T + \mathbf{K}_T^t = \mathbf{0}$. Thus

$$ \mathbf{K}_T = (\mathbf{D}_T^{-1} \mathbf{D}_N)^t. $$

Our equations for $\mathbf{C}_T$ and $\mathbf{K}_T$ show how the basic cycles and cuts associated with T can be deduced from the incidence matrix. We also have an algebraic proof of the following proposition.

Proposition 5.5　*Let T be a spanning tree of Γ and let a and b be edges of Γ such that $a \in T, b \notin T$. Then*

$$ b \in \mathrm{cut}(T, a) \Leftrightarrow a \in \mathrm{cyc}(T, b). $$

Proof　This result follows immediately from the definitions of $\mathbf{C}_T$ and $\mathbf{K}_T$, and the fact that $\mathbf{C}_T + \mathbf{K}_T^t = \mathbf{0}$. □

We end this chapter with a brief exposition of the solution of network equations; this application provided the stimulus for Kirchhoff's development of the foregoing theory in the middle of the nineteenth century.

An *electrical network* is a connected graph Γ (with an arbitrary orientation) which has certain physical characteristics, specified by two vectors in the edge-space of Γ. These vectors are the *current* vector $\mathbf{w}$ and the *voltage* vector $\mathbf{z}$. These vectors are related by a linear equation $\mathbf{z} = \mathbf{Mw} + \mathbf{n}$, where $\mathbf{M}$ is a diagonal matrix whose entries are the *conductances* of the edges, and $\mathbf{n}$ represents externally applied voltages. Further, $\mathbf{w}$ and $\mathbf{z}$ satisfy the equations

$$\mathbf{Dw} = \mathbf{0}, \quad \mathbf{C}^t\mathbf{z} = \mathbf{0},$$

which are known as *Kirchhoff's laws*. If we choose a spanning tree T in Γ, and partition $\mathbf{D}$ and $\mathbf{C}$ as before, then the same partition on $\mathbf{w}$ and $\mathbf{z}$ gives

$$\mathbf{w} = \begin{bmatrix} \mathbf{w}_T \\ \mathbf{w}_N \end{bmatrix}, \quad \mathbf{z} = \begin{bmatrix} \mathbf{z}_T \\ \mathbf{z}_N \end{bmatrix}.$$

Now, from $\mathbf{Dw} = \mathbf{0}$ we have $\mathbf{D}_T\mathbf{w}_T + \mathbf{D}_N\mathbf{w}_N = \mathbf{0}$, and since $\mathbf{C}_T = -\mathbf{D}_T^{-1}\mathbf{D}_N$ it follows that

$$\mathbf{w}_T = \mathbf{C}_T\mathbf{w}_N \quad \text{and} \quad \mathbf{w} = \mathbf{C}\mathbf{w}_N.$$

In other words, all the entries of the current vector are determined by the entries corresponding to edges not in T. Substituting in $\mathbf{z} = \mathbf{Mw} + \mathbf{n}$, and premultiplying by $\mathbf{C}^t$, we obtain

$$(\mathbf{C}^t\mathbf{MC})\mathbf{w}_N = -\mathbf{C}^t\mathbf{n}.$$

Since $\mathbf{C}^t\mathbf{MC}$ is a square matrix with size and rank both equal to $m-n+1$ it is invertible.

So this equation determines $\mathbf{w}_N$, and consequently both $\mathbf{w}$ (from $\mathbf{w} = \mathbf{Cw}_N$) and $\mathbf{z}$ (from $\mathbf{z} = \mathbf{Mw} + \mathbf{n}$) in turn. Thus we have a systematic method of solving network equations, which distinguishes clearly between the essential unknowns and the redundant ones.

Additional Results

5a *Total unimodularity*   A matrix is said to be *totally unimodular* if every square submatrix of it has determinant 0, 1, or -1; thus Proposition 5.3 states that $\mathbf{D}$ is totally unimodular. A generalisation of this result was proved by Heller and Tompkins (1956). They showed that if $\mathbf{M}$ is a matrix with elements 0, 1, or -1 such that every column contains at most two non-zero elements, then $\mathbf{M}$ is totally unimodular if and only if its rows can be partitioned into two disjoint parts satisfying:

(i) if a column has two non-zero elements with the same sign, then their rows are in different parts;

(ii) if a column has two non-zero elements with opposite signs, then their rows are in the same part.

5b *Integral solutions of LP problems*   Hoffman and Kruskal (1956) proved the following result. If **M** is a totally unimodular matrix and **b** is an integral vector then, for each objective function **c**, the linear programming problem (LP)

$$\text{maximise} \quad \mathbf{c}^t \mathbf{x} \quad \text{subject to} \quad \mathbf{Mx} \leq \mathbf{b}$$

has an optimal solution which is integral, provided that there is a finite solution.

Several optimization problems on graphs have LP formulations in which **M** is the incidence matrix, or a modified form of it. Among them are the maximum flow problem and the shortest path problem, the details of which are given in the standard text of Grötschel, Lovász and Schrijver (1988). Hoffman and Kruskal's theorem leads to 'integrality' results, such as the fact that if the capacities are integral then there is a maximum flow which is also integral.

5c *The unoriented incidence matrix*   As in Chapter 3, let **X** denote the matrix obtained from the incidence matrix **D** of Γ by replacing each entry ± 1 by $+1$. It follows from the result of Heller and Tompkins quoted in **5a** that Γ is bipartite if and only if **X** is totally unimodular. This was first observed by Egerváry (1931).

5d *The image of D again*   With the notation of **4g**, if ω is *integer-valued* and $S(\omega) = 0$ then there is an integer-valued ξ such that $D(\xi) = \omega$.

5e *The inverse of* $\mathbf{D}_T$ Let T be a spanning tree for Γ and let $\mathbf{D}_T$ denote the corresponding $(n-1) \times (n-1)$ matrix. Then $(\mathbf{D}_T^{-1})_{ij} = \pm 1$ if the edge e_i occurs in the unique path in T joining v_j to v_n. Otherwise $(\mathbf{D}_T^{-1})_{ij} = 0$.

5f *The Laplacian formulation of network equations*   For simplicity, consider the case of a network in which each edge has conductance 1. Then the network equations are

$$\mathbf{z} = \mathbf{w} + \mathbf{n}, \quad \mathbf{Dw} = 0, \quad \mathbf{C}^t \mathbf{z} = 0.$$

The last equation says that **z** is orthogonal to the cycle-subspace and so, by Definition 4.6, it belongs to the cut-subspace. It follows from **4a** that $\mathbf{z} = \mathbf{D}^t \phi$ for some *potential* ϕ in the vertex-space. Using the other two equations we obtain

$$\mathbf{D}\mathbf{D}^t \phi = \mathbf{D}\mathbf{n}; \quad \text{that is,} \quad \mathbf{Q}\phi = \eta,$$

where $\mathbf{Q}$ is the Laplacian matrix and η is a vector in which η_v is the current flowing into the network at the vertex v. In particular, defining

$$\eta_v^{xy} = \begin{cases} +1, & \text{if } v = x; \\ -1, & \text{if } v = y; \\ 0, & \text{otherwise;} \end{cases}$$

we see that the solution of the network equations when a current I enters at x and leaves at y is given by finding the potential satisfying $\mathbf{Q}\phi = I\eta^{xy}$.

5g *Existence and uniqueness of the solution; Thomson's principle*   Simple proofs of the results in the following paragraphs may be be found in a paper by Thomassen (1990). If x and y are vertices of a finite graph then there is a unique solution ϕ to the network equations for the case when a positive real-valued current I enters at x and leaves at y. The current vector $\mathbf{z} = \mathbf{D}^t \phi$ is the vector satisfying $\mathbf{D}\mathbf{z} = I\eta^{xy}$ for which the *power* $\|\mathbf{z}\|^2$ is a minimum. (This is known as *Thomson's principle*.)

5h *An explicit solution for the network equations*   Suppose that x and y are adjacent vertices of a connected graph Γ, and let κ denote the total number of spanning trees of Γ. (See Chapter 6 for more about κ.) For each spanning tree T of Γ send a current I/κ along the unique path in T from x to y. Then the current vector $\mathbf{z}$ which solves the network equations for a current I entering at x and leaving at y is the sum of these currents taken over all T. This result goes back to Kirchhoff (1847). For historical details and an algebraic proof, see Nerode and Shank (1961).

5i *The effective resistance*   For any two vertices x and y let ϕ be the potential satisfying $\mathbf{Q}\phi = I\eta^{xy}$. Following Ohm's law, the *effective resistance* from x to y is defined to be $(\phi_x - \phi_y)/I$. If x and y are adjacent vertices this is equal to κ_{xy}/κ, where κ_{xy} is the number of spanning trees which contain the edge $\{x, y\}$.

For example, it can be shown (see p. 39) that the number of spanning trees of the complete graph K_n is n^{n-2}; since each one contains $n - 1$ of the $n(n - 1)/2$ edges, there are $2n^{n-3}$ spanning trees containing a given edge. It follows that the effective resistance across an edge of K_n

is $2/n$. In general, if a graph has n vertices and m edges, and it is edge-transitive (see Chapter 15), then the effective resistance across an edge is $(n-1)/m$.

5j *Monotonicity results*   Let $R(x, y, \Gamma)$ denote the effective resistance of Γ from x to y. If Γ' is obtained from Γ by removing an edge (the *cutting* operation), then

$$R(x, y, \Gamma') \geq R(x, y, \Gamma).$$

The inequality is reversed if Γ' is obtained from Γ by identifying two vertices (the *shorting* operation). These results are known as *Rayleigh's monotonicity law*.

6

The tree-number

Several famous results in algebraic graph theory, including one of the oldest, are formulae for the numbers of spanning trees of certain graphs. Many formulae of this kind were given in the monograph written by Moon (1970). We shall show how such results can be derived from the Laplacian matrix $\mathbf{Q}$ introduced in Chapter 4.

Definition 6.1 The number of spanning trees of a graph Γ is its *tree-number*, denoted by $\kappa(\Gamma)$.

Of course, if Γ is disconnected, then $\kappa(\Gamma) = 0$. For the connected case, Theorem 6.3 below is a version of a formula for $\kappa(\Gamma)$ which has been discovered many times. We need a preparatory lemma concerning the matrix of cofactors (adjugate) of $\mathbf{Q}$.

Lemma 6.2 *Let $\mathbf{D}$ be the incidence matrix of a graph Γ, and let $\mathbf{Q} = \mathbf{DD}^t$ be the Laplacian matrix. Then the adjugate of $\mathbf{Q}$ is a multiple of $\mathbf{J}$.*

Proof Let n be the number of vertices of Γ. If Γ is disconnected, then

$$\text{rank } (\mathbf{Q}) = \text{rank } (\mathbf{D}) < n - 1,$$

and so every cofactor of $\mathbf{Q}$ is zero. That is, adj $\mathbf{Q} = \mathbf{0} = 0\mathbf{J}$.

If Γ is connected, then the ranks of $\mathbf{D}$ and $\mathbf{Q}$ are $n - 1$. Since

$$\mathbf{Q} \text{ adj } \mathbf{Q} = (\det \mathbf{Q})\mathbf{I} = \mathbf{0},$$

it follows that each column of adj $\mathbf{Q}$ belongs to the kernel of $\mathbf{Q}$. But this kernel is a one-dimensional space, spanned by $\mathbf{u} = [1, 1, \ldots, 1]^t$. Thus, each column of adj $\mathbf{Q}$ is a multiple of $\mathbf{u}$. Since $\mathbf{Q}$ is symmetric, so is adj $\mathbf{Q}$, and all the multipliers must be equal. Hence, adj $\mathbf{Q}$ is a multiple of $\mathbf{J}$. □

Theorem 6.3 *Every cofactor of $\mathbf{Q}$ is equal to the tree-number of Γ, that is,*

$$\text{adj } \mathbf{Q} = \kappa(\Gamma)\mathbf{J}.$$

Proof By Lemma 6.2, it is sufficient to show that one cofactor of $\mathbf{Q}$ is equal to $\kappa(\Gamma)$. Let $\mathbf{D}_0$ denote the matrix obtained from $\mathbf{D}$ by removing the last row; then det $\mathbf{D}_0\mathbf{D}_0^t$ is a cofactor of $\mathbf{Q}$. This determinant can be expanded by the Binet–Cauchy theorem (see *Theory of Matrices* by P. Lancaster (Academic Press) 1969, p. 38). The expansion is

$$\det(\mathbf{D}_0\mathbf{D}_0^t) = \sum_{|U|=n-1} \det(\mathbf{D}_U)\det(\mathbf{D}_U^t),$$

where $\mathbf{D}_U$ denotes the square submatrix of $\mathbf{D}_0$ whose $n-1$ columns correspond to the edges in a subset U of $E\Gamma$. Now, by Proposition 5.4, $\det \mathbf{D}_U$ is non-zero if and only if the subgraph $\langle U \rangle$ is a spanning tree for Γ, and then $\det \mathbf{D}_U$ takes the values ± 1. Since $\det \mathbf{D}_U^t = \det \mathbf{D}_U$, we have $\det(\mathbf{D}_0\mathbf{D}_0^t) = \kappa(\Gamma)$, and the result follows. □

For the complete graph K_n we have $\mathbf{Q} = n\mathbf{I} - \mathbf{J}$. A simple determinant manipulation on $n\mathbf{I} - \mathbf{J}$ with one row and column removed shows that $\kappa(K_n) = n^{n-2}$. This result was first obtained, for small values of n, by Cayley (1889).

We can dispense with the rather arbitrary procedure of removing one row and column from $\mathbf{Q}$, by means of the following result.

Proposition 6.4 (Temperley 1964) *The tree-number of a graph Γ with n vertices is given by the formula*

$$\kappa(\Gamma) = n^{-2}\det (\mathbf{J} + \mathbf{Q}).$$

Proof Since $n\mathbf{J} = \mathbf{J}^2$ and $\mathbf{J}\mathbf{Q} = \mathbf{0}$ we have the following equation:

$$(n\mathbf{I} - \mathbf{J})(\mathbf{J} + \mathbf{Q}) = n\mathbf{J} + n\mathbf{Q} - \mathbf{J}^2 - \mathbf{J}\mathbf{Q} = n\mathbf{Q}.$$

Thus, taking adjugates and using Theorem 6.3, we can argue as follows, where $\kappa = \kappa(\Gamma)$:

$$\text{adj } (\mathbf{J} + \mathbf{Q})\text{adj } (n\mathbf{I} - \mathbf{J}) = \text{adj } n\mathbf{Q},$$

$$\text{adj } (\mathbf{J} + \mathbf{Q})n^{n-2}\mathbf{J} = n^{n-1}\text{adj } \mathbf{Q},$$

$$\text{adj } (\mathbf{J} + \mathbf{Q})\mathbf{J} = n\kappa\mathbf{J},$$

$$(\mathbf{J} + \mathbf{Q}) \text{ adj } (\mathbf{J} + \mathbf{Q})\mathbf{J} = (\mathbf{J} + \mathbf{Q})n\kappa\mathbf{J},$$

$$\det (\mathbf{J} + \mathbf{Q})\mathbf{J} = n^2 \kappa \mathbf{J}.$$

It follows that $\det(\mathbf{J} + \mathbf{Q}) = n^2 \kappa$, as required. □

The next result uses the Laplacian spectrum introduced in **4e**.

Corollary 6.5   *Let* $0 \leq \mu_1 \leq \ldots \leq \mu_{n-1}$ *be the Laplacian spectrum of a graph* Γ *with* n *vertices. Then*

$$\kappa(\Gamma) = \frac{\mu_1 \mu_2 \cdots \mu_{n-1}}{n}.$$

If Γ *is connected and* k-*regular, and its spectrum is*

$$\text{Spec } \Gamma = \begin{pmatrix} k & \lambda_1 & \ldots & \lambda_{s-1} \\ 1 & m_1 & \ldots & m_{s-1} \end{pmatrix},$$

then

$$\kappa(\Gamma) = n^{-1} \prod_{r=1}^{s-1} (k - \lambda_r)^{m_r} = n^{-1} \chi'(\Gamma; k),$$

where χ' *denotes the derivative of the characteristic polynomial* χ.

Proof   Since $\mathbf{Q}$ and $\mathbf{J}$ commute, the eigenvalues of $\mathbf{J} + \mathbf{Q}$ are the sums of corresponding eigenvalues of $\mathbf{J}$ and $\mathbf{Q}$. The eigenvalues of $\mathbf{J}$ are $n, 0, 0, \ldots, 0$, so the eigenvalues of $\mathbf{J}+\mathbf{Q}$ are $n, \mu_1, \mu_2, \ldots, \mu_{n-1}$. Since the determinant is the product of the eigenvalues, the first formula follows.

In the case of a regular graph of degree k, an (ordinary) eigenvalue λ is $k - \mu$, where μ is a Laplacian eigenvalue. The result follows by collecting the eigenvalues according to their multiplicities, and recalling that $k - \lambda$ is a simple factor of χ in the connected case. □

Later in this book, when we have developed techniques for calculating the spectra of highly regular graphs, we shall be able to use this Corollary to write down the tree-numbers of many well-known families of graphs. For the moment, we shall consider applications of Corollary 6.5 in some simple, but important, cases. If Γ is a regular graph of degree k, then the characteristic polynomial of its line graph $L(\Gamma)$ is known in terms of that of Γ (Theorem 3.8). If Γ has n vertices and m edges, so that $2m = nk$, then we have

$$\kappa(L(\Gamma)) = m^{-1} \chi'(L(\Gamma); 2k - 2),$$

$$\kappa(\Gamma) = n^{-1} \chi'(\Gamma; k).$$

Differentiating the result of Theorem 3.8 and putting $\lambda = 2k - 2$, we get

$$\chi'(L(\Gamma); 2k - 2) = (2k)^{m-n} \chi'(\Gamma; k).$$

Hence we obtain the tree-number of Γ in terms of that of $L(\Gamma)$:

$$\kappa(L(\Gamma)) = 2^{m-n+1} k^{m-n} \kappa(\Gamma).$$

For example, the tree-number of the triangle graph $\Delta_t = L(K_t)$ is
$$\kappa(\Delta_t) = 2^{\frac{1}{2}(t^2-3t+2)}(t-1)^{\frac{1}{2}(t^2-3t-2)}t^{t-2}.$$

The *complete multipartite graph* $K_{a_1,a_2,\ldots,a_s}$ has a vertex-set which is partitioned into s parts $A_1, A_2, \ldots, A_s$, where $|A_i| = a_i$ $(1 \leq i \leq s)$; two vertices are joined by an edge if and only if they belong to different parts. In general this graph is not regular, but its complement (as defined in **3a**) consists of regular connected components. The tree-number of such graphs can be found by a modification of Proposition 6.4, due to Moon (1967). This is based on the properties of the characteristic function of the Laplacian matrix:
$$\sigma(\Gamma;\mu) = \det(\mu\mathbf{I} - \mathbf{Q}).$$

Proposition 6.6 (1) *If Γ is disconnected, then the σ function for Γ is the product of the σ functions for the components of Γ.*

(2) *If Γ is a k-regular graph, then $\sigma(\Gamma;\mu) = (-1)^n\chi(\Gamma; k-\mu)$, where χ is the characteristic polynomial of the adjacency matrix.*

(3) *If Γ^c is the complement of Γ, and Γ has n vertices, then*
$$\kappa(\Gamma) = n^{-2}\sigma(\Gamma^c; n).$$

Proof (1) This follows directly from the definition of σ.

(2) In the k-regular case, we have
$$\det(\mu\mathbf{I} - \mathbf{Q}) = \det(\mu\mathbf{I} - (k\mathbf{I} - \mathbf{A})) = (-1)^n\det((k-\mu)\mathbf{I} - \mathbf{A}),$$
whence the result.

(3) Let $\mathbf{Q}^c$ denote the Laplacian matrix for Γ^c, so that $\mathbf{Q}+\mathbf{Q}^c = n\mathbf{I}-\mathbf{J}$. Then using Proposition 6.4, we have
$$\kappa(\Gamma) = n^{-2}\det(\mathbf{J} + \mathbf{Q}) = n^{-2}\det(n\mathbf{I} - \mathbf{Q}^c) = n^{-2}\sigma(\Gamma^c; n).$$

$\square$

Consider the complete multipartite graph $K_{a_1,a_2,\ldots,a_s}$, where $a_1 + a_2 + \ldots + a_s = n$, the complement of which consists of s components isomorphic with $K_{a_1}, K_{a_2}, \ldots, K_{a_s}$. We know that $\chi(K_n; \lambda) = (\lambda+1)^{n-1}(\lambda - n + 1)$, and, using part (2) of Proposition 6.6, we obtain
$$\sigma(K_a;\mu) = (-1)^a\chi(K_a; a - 1 - \mu) = \mu(\mu - a)^{a-1}.$$

Consequently, applying parts (1) and (3) of Proposition 6.6,
$$\kappa(K_{a_1,a_2,\ldots,a_s}) = n^{-2}.(n)(n-a_1)^{a_1-1}\ldots(n)(n-a_s)^{a_s-1}$$
$$= n^{s-2}(n-a_1)^{a_1-1}\ldots(n-a_s)^{a_s-1}.$$

This result was originally found (by different means) by Austin (1960). We note the special cases:
$$\kappa(K_{a,b}) = a^{b-1}b^{a-1}, \quad \kappa(H_s) = 2^{2s-2}s^{s-1}(s-1)^s.$$

Additional Results

6a *A bound for the tree-number of a regular graph*   If Γ is a connected k-regular graph with n vertices, then applying the arithmetic–geometric mean inequality to the product formula in Corollary 6.5 we obtain

$$\kappa(\Gamma) \leq \frac{1}{n} \left(\frac{nk}{n-1} \right)^{n-1},$$

with equality if and only if $\Gamma = K_n$.

6b *More bounds for the tree-number*   Grimmett (1976) showed that the bound in **6a** can be extended to non-regular graphs. The result for any graph with m edges is

$$\kappa(\Gamma) \leq \frac{1}{n} \left(\frac{2m}{n-1} \right)^{n-1}.$$

This is clearly a generalisation of result **6a**, since $2m = nk$ in the k-regular case. Grone and Merris (1988) showed that if $\pi(\Gamma)$ is the product of the vertex-degrees then

$$\kappa(\Gamma) \leq \left(\frac{n}{n-1} \right)^{n-1} \left(\frac{\pi(\Gamma)}{2m} \right),$$

with equality if and only if $\Gamma = K_n$.

6c *A recursion for the tree-number*   For any (general) graph Γ, and any edge e which is not a loop, we define the graph $\Gamma^{(e)}$ to be the subgraph obtained by removing e, and $\Gamma_{(e)}$ to be the graph obtained from $\Gamma^{(e)}$ by identifying the vertices of e. Note that even if Γ itself is a graph (rather than a general graph) this process may produce a general graph. We have

$$\kappa(\Gamma) = \kappa(\Gamma^{(e)}) + \kappa(\Gamma_{(e)}).$$

6d *Tree-number of a Möbius ladder*   The tree-number of the Möbius ladder M_h defined in **3e** may be computed in two ways. Using the spectral formula **6.5** we obtain

$$\kappa(M_h) = \frac{1}{2h} \prod_{j=1}^{2h-1} \left(3 - (-1)^j - 2\cos\frac{\pi j}{h} \right).$$

An alternative is to use **6c** to obtain a recursion formula. Sedlacek (1970) used this method to obtain

$$\kappa(M_h) = \frac{h}{2}[(2 + \sqrt{3})^h + (2 - \sqrt{3})^h] + h.$$

The recursive method was discussed in greater generality by Biggs, Damerell and Sands (1972); see **9i**.

6e *Almost-complete graphs* Let Γ be a graph constructed by removing q disjoint edges from K_n where $n \geq 2q$. Then

$$\kappa(\Gamma) = n^{n-2}\left(1 - \frac{2}{n}\right)^q.$$

In particular, taking $n = 2q$, we have the formula for the tree-number of H_q.

6f *Tree-numbers of planar duals* Let Γ and Γ^* be dual planar graphs (as defined in **4f**) and let $\mathbf{D}$ and $\mathbf{D}^*$ be the corresponding incidence matrices. Suppose that Γ has n vertices, Γ^* has n^* vertices, and $|E\Gamma| = |E\Gamma^*| = m$; then $(n-1) + (n^*-1) = m$. If $\mathbf{D}_U$ is a square submatrix of $\mathbf{D}$, whose $n-1$ columns correspond to the edges of a subset U of $E\Gamma$, and U^* denotes the complementary subset of $E\Gamma^* = E\Gamma$, then $\mathbf{D}_U$ is non-singular if and only if $\mathbf{D}_U^*$ is non-singular. Consequently,

$$\kappa(\Gamma) = \kappa(\Gamma^*).$$

6g *The octahedron and the cube* The *octahedron graph* is $H_3 = K_{2,2,2}$; it is planar and the *cube graph* Q_3 is its dual. We have

$$\text{Spec } H_3 = \begin{pmatrix} 4 & 0 & -2 \\ 1 & 3 & 2 \end{pmatrix}, \quad \text{Spec } Q_3 = \begin{pmatrix} 3 & 1 & -1 & -3 \\ 1 & 3 & 3 & 1 \end{pmatrix}.$$

Hence $\kappa(H_3) = \kappa(Q_3) = 384$, in agreement with **6e**.

6h *The σ function of the complement* From the equation $\mathbf{Q} + \mathbf{Q}^c = n\mathbf{I} - \mathbf{J}$, we obtain

$$\mu\mathbf{I} - \mathbf{Q}^c = [(n-\mu)^{-1}\mathbf{J} - \mathbf{I}][(n-\mu)\mathbf{I} - \mathbf{Q}].$$

Taking determinants we have

$$(n-\mu)\,\sigma(\Gamma^c; \mu) = (-1)^{n-1}\mu\,\sigma(\Gamma; n-\mu).$$

6i *Spectral characterization of complete multipartite graphs* The complete multipartite graphs defined on page 41 are the only connected graphs for which the second largest eigenvalue λ_1 is not positive (Smith 1970).

7

Determinant expansions

In this chapter we shall investigate the characteristic polynomial χ, and the polynomial σ introduced in Chapter 6, by means of determinant expansions. We begin by considering the determinant of the adjacency matrix $\mathbf{A}$ of a graph Γ. We suppose, as before, that $V\Gamma = \{v_1, v_2, \ldots, v_n\}$ and that the rows and columns of $\mathbf{A}$ are labelled to conform with this notation. The expansion which is useful here is the usual definition of a determinant: if $\mathbf{A} = (a_{ij})$, then

$$\det \mathbf{A} = \sum \operatorname{sgn}(\pi) a_{1,\pi 1} a_{2,\pi 2} \ldots a_{n,\pi n},$$

where the summation is over all permutations π of the set $\{1, 2, \ldots, n\}$.

In order to express the quantities which appear in the above expansion in graph-theoretical terms, it is helpful to introduce a new definition.

Definition 7.1 An *elementary* graph is a simple graph, each component of which is regular and has degree 1 or 2. In other words, each component is a single edge (K_2) or a cycle (C_r). A *spanning elementary subgraph* of Γ is an elementary subgraph which contains all vertices of Γ.

We observe that the co-rank of an elementary graph is just the number of its components which are cycles.

Proposition 7.2 (Harary 1962) *Let $\mathbf{A}$ be the adjacency matrix of a graph Γ. Then*

$$\det \mathbf{A} = \sum (-1)^{r(\Lambda)} 2^{s(\Lambda)},$$

where the summation is over all spanning elementary subgraphs Λ of Γ.

Proof   Consider a term $\text{sgn}(\pi)a_{1,\pi 1}a_{2,\pi 2}\cdots a_{n,\pi n}$ in the expansion of $\det \mathbf{A}$. This term vanishes if, for some $i \in \{1, 2, \ldots, n\}$, $a_{i,\pi i} = 0$; that is, if $\{v_i, v_{\pi i}\}$ is not an edge of Γ. In particular, the term vanishes if π fixes any symbol. Thus, if the term corresponding to a permutation π is non-zero, then π can be expressed uniquely as the composition of disjoint cycles of length at least two. Each cycle (ij) of length two corresponds to the factors $a_{ij}a_{ji}$, and signifies a single edge $\{v_i, v_j\}$ in Γ. Each cycle $(pqr\ldots t)$ of length greater than two corresponds to the factors $a_{pq}a_{qr}\ldots a_{tp}$, and signifies a cycle $\{v_p, v_q, \ldots, v_t\}$ in Γ. Consequently, each non-vanishing term in the determinant expansion gives rise to an elementary subgraph Λ of Γ, with $V\Lambda = V\Gamma$.

The sign of a permutation π is $(-1)^{N_e}$, where N_e is the number of even cycles in π. If there are c_l cycles of length l, then the equation $\Sigma l c_l = n$ shows that the number N_o of odd cycles is congruent to n modulo 2. Hence,

$$r(\Lambda) = n - (N_o + N_e) \equiv N_e \pmod 2,$$

so the sign of π is equal to $(-1)^{r(\Lambda)}$.

Each elementary subgraph Λ with n vertices gives rise to several permutations π for which the corresponding term in the determinant expansion does not vanish. The number of such π arising from a given Λ is $2^{s(\Lambda)}$, since for each cycle-component in Λ there are two ways of choosing the corresponding cycle in π. Thus each Λ contributes $(-1)^{r(\Lambda)}2^{s(\Lambda)}$ to the determinant, and we have the result. $\Box$

For example, in the complete graph K_4 there are just two kinds of elementary subgraph with four vertices: pairs of disjoint edges (for which $r = 2$ and $s = 0$) and 4-cycles (for which $r = 3$ and $s = 1$. There are three subgraphs of each kind so we have

$$\det \mathbf{A}(K_4) = 3(-1)^2 2^0 + 3(-1)^3 2^1 = -3.$$

At the beginning of this book we obtained a description of the first few coefficients of the characteristic polynomial of Γ, in terms of some small subgraphs of Γ (Proposition 2.3). We shall now extend that result to all the coefficients. We shall suppose, as before, that

$$\chi(\Gamma; \lambda) = \lambda^n + c_1\lambda^{n-1} + c_2\lambda^{n-2} + \ldots + c_n.$$

Proposition 7.3   *The coefficients of the characteristic polynomial are given by*

$$(-1)^i c_i = \sum (-1)^{r(\Lambda)}2^{s(\Lambda)},$$

where the summation is over all elementary subgraphs Λ of Γ with i vertices.

Proof   The number $(-1)^i c_i$ is the sum of all principal minors of $\mathbf{A}$ with i rows and columns. Each such minor is the determinant of the adjacency matrix of an induced subgraph of Γ with i vertices. Any elementary subgraph with i vertices is contained in precisely one of these induced subgraphs, and so, by applying Proposition 7.2 to each minor, we obtain the required result. $\square$

The only elementary graphs with fewer than four vertices are: K_2 (an edge), and C_3 (a triangle). Thus, we can immediately regain the results of Proposition 2.3 from the general formula of Proposition 7.3. We can also use Proposition 7.3 to derive explicit expressions for the other coefficients, for example, c_4. Since the only elementary graphs with four vertices are the cycle graph C_4 and the graph having two disjoint edges, it follows that

$$c_4 = n_a - 2n_b,$$

where n_a is the number of pairs of disjoint edges in Γ, and n_b is the number of 4-cycles in Γ. (See **7i**.)

As well as giving explicit expressions for the coefficients of the characteristic polynomial, Proposition 7.3 throws some light on the problem of cospectral graphs (**2f**). The fact that elementary subgraphs are rather loosely related to the structure of a graph helps to explain why there are many pairs of non-isomorphic graphs having the same spectrum. This is particularly so in the case of trees (see **7b** and **7c**).

We now turn to an expansion of the characteristic function of the Laplacian matrix

$$\sigma(\Gamma; \mu) = \det(\mu \mathbf{I} - \mathbf{Q}).$$

Although the Laplacian matrix $\mathbf{Q}$ differs from $-\mathbf{A}$ only in its diagonal entries, the ideas involved in this expansion are quite different from those which we have used to investigate the characteristic polynomial of $\mathbf{A}$. One reason for this is that a principal submatrix of $\mathbf{Q}$ is (in general) not the Laplacian matrix of an induced subgraph of Γ (the diagonal entries give the degrees in Γ, rather than in the subgraph).

We shall write

$$\sigma(\Gamma; \mu) = \det(\mu \mathbf{I} - \mathbf{Q}) = \mu^n + q_1 \mu^{n-1} + \ldots + q_{n-1}\mu + q_n.$$

The coefficient $(-1)^i q_i$ is the sum of the principal minors of $\mathbf{Q}$ which have i rows and columns. Using results from Chapter 6 and some simple observations we obtain

$$q_1 = -2|E\Gamma|, \quad q_{n-1} = (-1)^{n-1}n\kappa(\Gamma), \quad q_n = 0.$$

We shall find a general expression for q_i which subsumes these results. The method is based on the expansion of a principal minor of $\mathbf{Q} = \mathbf{DD}^t$ by means of the Binet–Cauchy theorem, as in the proof of Theorem 6.3.

Let X be a non-empty subset of the vertex-set of Γ, and Y a non-empty subset of the edge-set of Γ. We denote by $\mathbf{D}(X,Y)$ the submatrix of the incidence matrix $\mathbf{D}$ of Γ defined by the rows corresponding to vertices in X and the columns corresponding to edges in Y. The following lemma amplifies the results of Propositions 5.3 and 5.4.

Lemma 7.4 *Let X and Y be as above, with $|X| = |Y|$, and let V_0 denote the vertex-set of the subgraph $\langle Y \rangle$. Then $\mathbf{D}(X,Y)$ is invertible if and only if the following conditions are satisfied:*

(1) *X is a subset of V_0;*

(2) *$\langle Y \rangle$ contains no cycles;*

(3) *$V_0 \setminus X$ contains precisely one vertex from each component of $\langle Y \rangle$.*

Proof Suppose that $\mathbf{D}(X,Y)$ is invertible. If X were not a subset of V_0, then $\mathbf{D}(X,Y)$ would contain a row of zeros and would not be invertible; hence condition (1) holds. The matrix $\mathbf{D}(V_0,Y)$ is the incidence matrix of $\langle Y \rangle$, and if $\langle Y \rangle$ contains a cycle then $\mathbf{D}(V_0,Y)\mathbf{z} = 0$ for the vector $\mathbf{z}$ representing this cycle. Consequently $\mathbf{D}(X,Y)\mathbf{z} = \mathbf{0}$ and $\mathbf{D}(X,Y)$ is not invertible. Thus condition (2) holds. It follows that the co-rank of $\langle Y \rangle$ is zero; that is,

$$|Y| - |V_0| + c = 0,$$

where c is the number of components of $\langle Y \rangle$. Since $|X| = |Y|$ we have $|V_0 \setminus X| = c$. If X contained all the vertices from some component of $\langle Y \rangle$, then the corresponding rows of $\mathbf{D}(X,Y)$ would sum to 0, and $\mathbf{D}(X,Y)$ would not be invertible. Thus $V_0 \setminus X$ contains some vertices from each component of $\langle Y \rangle$, and since $|V_0 \setminus X| = c$, it must contain precisely one vertex from each component, and condition (3) is verified.

The converse is proved by reversing the argument. $\square$

A graph Φ whose co-rank is zero is a *forest*; it is the union of components each of which is a tree. We shall use the symbol $p(\Phi)$ to denote the product of the numbers of vertices in the components of Φ. In particular, if Φ is connected it is a tree, and we have $p(\Phi) = |V\Phi|$.

Theorem 7.5 *The coefficients q_i of the polynomial $\sigma(\Gamma;\mu)$ are given by the formula*

$$(-1)^i q_i = \sum p(\Phi) \quad (1 \le i \le n),$$

where the summation is over all sub-forests Φ of Γ which have i edges.

Proof   Let $\mathbf{Q}_X$ denote the principal submatrix of $\mathbf{Q}$ whose rows and columns correspond to the vertices in a subset X of $V\Gamma$. Then $q_i = \sum \det \mathbf{Q}_X$, where the summation is over all X with $|X| = i$. Using the notation of Lemma 7.4 and the fact that $\mathbf{Q} = \mathbf{DD}^t$, it follows from the Binet–Cauchy theorem that

$$\det \mathbf{Q}_X = \sum \det \mathbf{D}(X,Y) \det \mathbf{D}(X,Y)^t = \sum (\det \mathbf{D}(X,Y))^2.$$

This summation is over all subsets Y of $E\Gamma$ with $|Y| = |X| = i$. Thus,

$$q_i = \sum_{X,Y} (\det \mathbf{D}(X,Y))^2.$$

By Proposition 5.3, $(\det \mathbf{D}(X,Y))^2$ is either 0 or 1, and it takes the value 1 if and only if the three conditions of Lemma 7.4 hold. For each forest $\Phi = \langle Y \rangle$ there are $p(\Phi)$ ways of omitting one vertex from each component of Φ, and consequently there are $p(\Phi)$ summands equal to 1 in the expression for q_i. This is the result.

Corollary 7.6   *The tree-number of a graph Γ is given by the formula*

$$\kappa(\Gamma) = n^{n-2} \sum p(\Phi)(-n)^{-|E\Phi|},$$

where the summation is over all forests Φ which are subgraphs of the complement of Γ.

Proof   The result of Proposition 6.6, part (3), expresses $\kappa(\Gamma)$ in terms of the σ function of Γ^c. The stated result follows from the formula of Theorem 7.5 for the coefficients of σ.   □

This formula can be useful when the complement of Γ is relatively small; examples of this situation are given in **6e** and **7d**. In the case of a regular graph Γ, the relationship between σ and χ leads to an interesting consequence of Theorem 7.5.

Proposition 7.7   *Let Γ be a regular graph of degree k, and let $\chi^{(i)}$ ($0 \leq i \leq n$) denote the ith derivative of the characteristic polynomial of Γ. Then*

$$\chi^{(i)}(\Gamma; k) = i! \sum p(\Phi),$$

where the summation is over all forests Φ which are subgraphs of Γ with $|E\Phi| = n - i$.

Proof   From part (2) of Proposition 6.6, we have

$$\sigma(\Gamma; \mu) = (-1)^{n-1} \chi(\Gamma; k - \mu).$$

The Taylor expansion of χ at the value k can be written in the form

$$\chi(\Gamma; k - \mu) = \sum_{i=0}^{n} \chi^{(i)}(\Gamma; k) \frac{(-\mu)^i}{i!}.$$

Comparing this with $\sigma(\Gamma; \mu) = \sum q_{n-i} \mu^i$, we have the result. $\qquad \square$

We notice that the case $i = 1$ of Proposition 7.7 gives

$$\chi'(\Gamma; k) = (-1)^{n-1} q_{n-1} = n\kappa(\Gamma),$$

which is just the formula given in Corollary 6.5.

Additional Results

7a *Odd cycles* (Sachs 1964)   Let $\chi(\Gamma; \lambda) = \sum c_{n-i} \lambda^i$ and suppose

$$c_3 = c_5 = \ldots = c_{2r-1} = 0, \quad c_{2r+1} \neq 0.$$

Then the shortest odd cycle in Γ has length $2r + 1$, and there are $-c_{2r+1}/2$ such cycles.

7b *The characteristic polynomial of a tree*   Suppose that $\sum c_i \lambda^{n-i}$ is the characteristic polynomial of a tree with n vertices. Then the odd coefficients c_{2r+1} are zero, and the even coefficients c_{2r} are given by the rule that $(-1)^r c_{2r}$ is the number of ways of choosing r disjoint edges in the tree.

7c *Cospectral trees*   The result **7b** facilitates the construction of pairs of cospectral trees. For example, there are two different trees with eight vertices and characteristic polynomial $\lambda^8 - 7\lambda^6 + 10\lambda^4$. Schwenk (1973) proved that if we select a tree T with n vertices, all such trees being equally likely, then the probability that T belongs to a cospectral pair tends to 1 as n tends to infinity.

7d *The σ function of a star graph*   A *star graph* is a complete bipartite graph $K_{1,b}$. For such a graph we can calculate σ explicitly from the formula of Theorem 7.5: the result is

$$\sigma(K_{1,b}; \mu) = \mu(\mu - b - 1)(\mu - 1)^{b-1}.$$

Consequently if Γ is the graph obtained by removing a star $K_{1,b}$ from K_n, where $n > b + 1$, we have

$$\kappa(\Gamma) = n^{n-2} \left(1 - \frac{1}{n}\right)^{b-1} \left(1 - \frac{b+1}{n}\right).$$

7e *Complete matchings, Hamiltonian cycles, and the determinant formula*   We may write the formula for det $\mathbf{A}$ as

$$\det \mathbf{A} = \sum f(r,s)(-1)^r 2^s,$$

where $f(r,s)$ is the number of spanning elementary subgraphs with rank r and co-rank s. Two terms in this formula have special significance. The number $f(n/2,0)$ is the number of disjoint edges which cover all the vertices – the *complete matchings*. The number $f(n-1,1)$ is the number of spanning elementary subgraphs which are connected, that is, the number of single cycles which cover all the vertices – the *Hamiltonian cycles*.

7f *Reconstruction – Kelly's lemma*   For each vertex $v \in V\Gamma$ let Γ_v denote the induced subgraph $\langle V\Gamma \setminus v \rangle$. The *deck* of Γ is the set of (unlabelled) induced subgraphs $\{\Gamma_v \mid v \in V\Gamma\}$. The graph is said to be *reconstructible* if every graph with the same deck as Γ is isomorphic to Γ. The *reconstruction conjecture* is that every graph with at least three vertices is reconstructible.

A function defined on graphs is said to be reconstructible if it takes the same value on all graphs with the same deck. For any graphs Γ and Δ, let $n(\Gamma, \Delta)$ be the number of subgraphs of Γ which are isomorphic to Δ. Standard double counting arguments lead to the formula

$$n(\Gamma, \Delta) = \frac{1}{|V\Gamma| - |V\Delta|} \sum_{v \in V\Gamma} n(\Gamma_v, \Delta).$$

From this formula it follows that $n(\Gamma, \Delta)$ is reconstructible whenever $|V\Delta| < |V\Gamma|$ (Kelly 1957).

7g *Reconstruction – Kocay's Lemma*   A sequence of graphs

$$\mathcal{F} = (\Phi_1, \Phi_2, \ldots, \Phi_l)$$

is said to be a *cover* of the graph Γ if there are subgraphs Δ_i of Γ such that Δ_i is isomorphic to Φ_i $(1 \le i \le l)$ and the union of the subgraphs is Γ. The number of covers of Γ by $\mathcal{F}$ is denoted by $c(\Gamma, \mathcal{F})$. Kocay (1981) proved that, provided all the members of the sequence $\mathcal{F}$ have fewer vertices than Γ, the function

$$\sum_X n(\Gamma, X) c(X, \mathcal{F})$$

is reconstructible, where the sum is taken over all isomorphism classes of graphs X such that $|VX| = |V\Gamma|$.

7h *The reconstructibility of the characteristic polynomial* Using the lemmas of Kelly and Kocay, and the formula in Proposition 7.3, it can be shown that the coefficients c_i of the characteristic polynomial are reconstructible. In particular $(-1)^n c_n = \det \mathbf{A}$ is reconstructible. These results were first established by Tutte (1979) using a different method. His proof, and that using Kocay's lemma (as given by Bondy (1991)) both depend on showing that the number of Hamiltonian cycles, denoted by $f(n-1,1)$ in **7e**, is reconstructible.

7i *Angles and the number of 4-cycles* The number of 3-cycles in a graph is determined by the spectrum (see **2h**), but the number of 4-cycles is not, except in special cases, such as when the graph is regular. However, the number of 4-cycles is determined by the spectrum and the *angles*, which are defined as follows. Let $\sum \lambda_a \mathbf{E}_a$ be the spectral decomposition of the adjacency matrix $\mathbf{A}$, as defined in **2j**, and let $\mathbf{e}_1, \mathbf{e}_2, \ldots, \mathbf{e}_n$ be the standard orthonormal basis for Euclidean n-space. Then the angles are the numbers

$$\alpha_{ij} = \mathbf{E}_i \mathbf{e}_j \quad (1 \le i, j \le n).$$

More about this construction, and an explicit formula for the number of 4-cycles, can be found in a paper by Cvetković and Rowlinson (1988).

7j *The Shannon capacity of a graph* Let $\Gamma\Lambda$ denote the product of graphs Γ and Λ obtained by taking the vertex-set to be Cartesian product of their vertex-sets, and defining two distinct vertices to be adjacent if both coordinates are equal or adjacent. Let Γ^r denote the product of r copies of Γ, and let $\alpha(\Gamma^r)$ denote the maximum number of mutually non-adjacent vertices of Γ^r. A construction in coding theory due to Shannon involves the quantity

$$\Theta(\Gamma) = \lim_{r \to \infty} (\alpha(\Gamma^r))^{1/r},$$

and this is known as the *Shannon capacity* of Γ. Since $\alpha(\Gamma)^r \le \alpha(\Gamma^r)$, it follows that $\alpha(\Gamma) \le \Theta(\Gamma)$, but in general equality does not hold. Lovász (1979) showed that $\Theta(\Gamma)$ is bounded above by the largest eigenvalue of any real symmetric matrix $\mathbf{C}$ for which $c_{ij} = 1$ whenever v_i and v_j are not adjacent. In particular, this yields the result $\Theta(C_5) = \sqrt{5}$.

8

Vertex-partitions and the spectrum

One of the oldest problems in graph theory is the vertex-colouring problem, which involves the assignment of colours to the vertices in such a way that adjacent vertices have different colours. This can be interpreted as a problem about a special kind of partition of the vertex-set, as described in the first definition below. In this chapter we shall apply spectral techniques to the vertex-colouring problem, using inequalities involving the eigenvalues of a graph. Similar techniques can also be applied to other problems about vertex-partitions, and some of these are mentioned in the Additional Results at the end of the chapter.

Definition 8.1 A *colour-partition* of a general graph Γ is a partition of $V\Gamma$ into subsets, called *colour-classes*,
$$V\Gamma = V_1 \cup V_2 \cup \ldots \cup V_l,$$
such that each V_i $(1 \le i \le l)$ contains no pair of adjacent vertices. In other words, the induced subgraphs $\langle V_i \rangle$ have no edges. The *chromatic number* of Γ, written $\nu(\Gamma)$, is the least natural number l for which such a partition is possible.

We define a *vertex-colouring* of Γ to be an assignment of colours to the vertices, with the property that adjacent vertices have different colours, so clearly, a vertex-colouring in which l colours are used gives rise to a colour-partition with l colour-classes.

We note that if Γ has a loop, then it has a self-adjacent vertex, and consequently no colour-partitions. Also, if Γ has several edges joining the same pair of vertices then only one of these edges is relevant to the definition of a colour-partition, since the definition depends only on whether vertices are adjacent or not. Thus we can continue, for the moment, to deal with strict graphs. However, this is allowable only for the purposes of the present chapter; some of the constructions used in Part Two require the introduction of general graphs.

If $\nu(\Gamma) = 1$, then Γ has no edges. If $\nu(\Gamma) = 2$ then Γ is a bipartite graph, as defined in **2c**. Since a cycle of odd length cannot be coloured with two colours, it follows that a bipartite graph contains no odd cycles. This observation leads to another proof of the result established in **2c**.

Proposition 8.2   *Suppose the bipartite graph Γ has an eigenvalue λ of multiplicity $m(\lambda)$. Then $-\lambda$ is also an eigenvalue of Γ, and $m(-\lambda) = m(\lambda)$.*

Proof   The formula of Proposition 7.3 expresses the characteristic polynomial of a graph Γ in terms of the elementary subgraphs of Γ. If Γ is bipartite then Γ has no odd cycles, and consequently no elementary subgraphs with an odd number of vertices. It follows that the characteristic polynomial of Γ has the form

$$\chi(\Gamma; z) = z^n + c_2 z^{n-2} + c_4 z^{n-4} + \ldots = z^\delta p(z^2),$$

where $\delta = 0$ or 1, and p is a polynomial function. Thus the eigenvalues, which are the zeros of χ, have the required property.   □

The spectrum of the complete bipartite graph $K_{a,b}$ can be found in the following manner. We may suppose that the vertices of $K_{a,b}$ are labelled in such a way that its adjacency matrix is

$$\mathbf{A} = \begin{bmatrix} \mathbf{0} & \mathbf{J} \\ \mathbf{J}^t & \mathbf{0} \end{bmatrix},$$

where $\mathbf{J}$ is the $a \times b$ matrix having all entries +1. The matrix $\mathbf{A}$ has just two linearly independent rows, and so its rank is 2. Consequently, 0 is an eigenvalue of $\mathbf{A}$ with multiplicity $a + b - 2$. The characteristic polynomial is thus of the form $z^{a+b-2}(z^2 + c_2)$. By Proposition 2.3, $-c_2$ is equal the number of edges of $K_{a,b}$, that is, ab. Hence

$$\text{Spec } K_{a,b} = \begin{pmatrix} \sqrt{(ab)} & 0 & -\sqrt{(ab)} \\ 1 & a+b-2 & 1 \end{pmatrix}.$$

This example illustrates the fact (Proposition 8.2) that the spectrum of a bipartite graph is symmetrical with respect to the origin. Indeed, the converse of this result is also true [CvDS, p. 87]. But if $\nu(\Gamma) > 2$ the spectrum of Γ does not have a distinctive property, as it does in the bipartite case. However, as we shall see, it is possible to make useful deductions about the chromatic number from a knowledge of the maximum and minimum eigenvalues of Γ.

For any real symmetric matrix $\mathbf{M}$, we shall denote the maximum and minimum eigenvalues of $\mathbf{M}$ by $\lambda_{\max}(\mathbf{M})$ and $\lambda_{\min}(\mathbf{M})$. If $\mathbf{M}$ is the adjacency matrix of a graph Γ we shall also use the notation $\lambda_{\max}(\Gamma)$ and $\lambda_{\min}(\Gamma)$. It follows from Proposition 8.2 that, for a bipartite graph Γ, we have $\lambda_{\min}(\Gamma) = -\lambda_{\max}(\Gamma)$.

We need a useful technique from matrix theory. Let $(\mathbf{x}, \mathbf{y})$ denote the inner product of the column vectors $\mathbf{x}$, $\mathbf{y}$. For any real $n \times n$ symmetric matrix $\mathbf{X}$, and any real non-zero $n \times 1$ column vector $\mathbf{z}$, the number $(\mathbf{z}, \mathbf{X}\mathbf{z})/(\mathbf{z}, \mathbf{z})$ is known as the *Rayleigh quotient*, and written $R(\mathbf{X}; \mathbf{z})$. In matrix theory it is proved that

$$\lambda_{\max}(\mathbf{X}) \geq R(\mathbf{X}; \mathbf{z}) \geq \lambda_{\min}(\mathbf{X}) \quad \text{for all} \quad \mathbf{z} \neq \mathbf{0},$$

a result which has important applications in spectral graph theory.

Proposition 8.3   (1) *If Λ is an induced subgraph of Γ, then*

$$\lambda_{\max}(\Lambda) \leq \lambda_{\max}(\Gamma); \quad \lambda_{\min}(\Lambda) \geq \lambda_{\min}(\Gamma).$$

(2) *If the greatest and least degrees among the vertices of Γ are $k_{\max}(\Gamma)$ and $k_{\min}(\Gamma)$, and the average degree is $k_{\text{ave}}(\Gamma)$, then*

$$k_{\max}(\Gamma) \geq \lambda_{\max}(\Gamma) \geq k_{\text{ave}}(\Gamma) \geq k_{\min}(\Gamma).$$

Proof   (1) We may suppose that the vertices of Γ are labelled so that the adjacency matrix $\mathbf{A}$ of Γ has a leading principal submatrix $\mathbf{A}_0$, which is the adjacency matrix of Λ. Let $\mathbf{z}_0$ be chosen such that $\mathbf{A}_0\mathbf{z}_0 = \lambda_{\max}(\mathbf{A}_0)\mathbf{z}_0$ and $(\mathbf{z}_0, \mathbf{z}_0) = 1$. Further, let $\mathbf{z}$ be the column vector with $|V\Gamma|$ rows formed by adjoining zero entries to $\mathbf{z}_0$. Then

$$\lambda_{\max}(\mathbf{A}_0) = R(\mathbf{A}_0; \mathbf{z}_0) = R(\mathbf{A}; \mathbf{z}) \leq \lambda_{\max}(\mathbf{A}).$$

That is, $\lambda_{\max}(\Lambda) \leq \lambda_{\max}(\Gamma)$. The other inequality is proved similarly.

(2) Let $\mathbf{u}$ be the column vector each of whose entries is $+1$. Then, if $n = |V\Gamma|$ and $k^{(i)}$ is the degree of the vertex v_i, we have

$$R(\mathbf{A}; \mathbf{u}) = \frac{1}{n} \sum_{i,j} a_{ij} = \frac{1}{n} \sum_i k^{(i)} = k_{\text{ave}}(\Gamma).$$

The Rayleigh quotient $R(\mathbf{A}; \mathbf{u})$ is at most $\lambda_{\max}(\mathbf{A})$, that is $\lambda_{\max}(\Gamma)$, and it is clear that the average degree is not less than the minimum degree. Hence

$$\lambda_{\max}(\Gamma) \geq k_{\text{ave}}(\Gamma) \geq k_{\min}(\Gamma).$$

Finally, let $\mathbf{x}$ be an eigenvector corresponding to the eigenvalue $\lambda_0 = \lambda_{\max}(\Gamma)$, and let x_j be a largest positive entry of $\mathbf{x}$. By an argument similar to that used in Proposition 3.1, we have

$$\lambda_0 x_j = (\lambda_0 \mathbf{x})_j = \Sigma' x_i \leq k^{(j)} x_j \leq k_{\max}(\Gamma) x_j,$$

where the sum Σ' is taken over the vertices v_i adjacent to v_j. Thus $\lambda_0 \leq k_{\max}(\Gamma)$. □

We shall now bound the chromatic number of Γ in terms of $\lambda_{\max}(\Gamma)$ and $\lambda_{\min}(\Gamma)$. A graph Γ is *l-critical* if $\nu(\Gamma) = l$, and for all induced subgraphs $\Lambda \neq \Gamma$ we have $\nu(\Lambda) < l$.

Lemma 8.4 *Suppose Γ is a graph with chromatic number $l \geq 2$. Then Γ has an l-critical induced subgraph Λ, and every vertex of Λ has degree at least $l - 1$ in Λ.*

Proof The set of all induced subgraphs of Γ is non-empty and contains some graphs (for example, Γ itself) whose chromatic number is l, and also some graphs (for example, those with one vertex) whose chromatic number is not l. Let Λ be an induced subgraph whose chromatic number is l, and which is minimal with respect to the number of vertices; then clearly Λ is l-critical. If v is any vertex of Λ, then $\langle V\Lambda \setminus v \rangle$ is an induced subgraph of Λ and has a vertex-colouring with $l-1$ colours. If the degree of v in Λ were less than $l - 1$, then we could extend this vertex-colouring to Λ, contradicting the fact that $\nu(\Lambda) = l$. Thus the degree of v is at least $l - 1$. □

Proposition 8.5 (Wilf 1967) *For any graph Γ we have*
$$\nu(\Gamma) \leq 1 + \lambda_{\max}(\Gamma).$$

Proof It follows from Lemma 8.4 that there is an induced subgraph Λ of Γ such that $\nu(\Lambda) = \nu(\Gamma)$, and $k_{\min}(\Lambda) \geq \nu(\Gamma) - 1$. Thus, using the inequalities of Proposition 8.3, we have
$$\nu(\Gamma) \leq 1 + k_{\min}(\Lambda) \leq 1 + \lambda_{\max}(\Lambda) \leq 1 + \lambda_{\max}(\Gamma).$$

□

Wilf's bound may be compared with the simple bound $\nu \leq 1 + k_{\max}$, which is proved by an obvious argument. There is also a nontrivial refinement of the simple bound, known as Brooks's theorem: $\nu \leq k_{\max}$ unless Γ is a complete graph or an odd cycle. For example, for the complete bipartite graph $K_{a,b}$ we have
$$k_{\max}(K_{a,b}) = \max(a,b), \quad \lambda_{\max}(K_{a,b}) = \sqrt{(ab)}.$$

When a is large in comparison with b the second number is much smaller than the first, but it is still a poor bound for the chromatic number, 2.

Our next major result is complementary to the previous one, in that it provides a lower bound for the chromatic number. We require a preliminary lemma and a corollary.

Lemma 8.6   *Let* $\mathbf{X}$ *be a real symmetric matrix, partitioned in the form*
$$\mathbf{X} = \begin{bmatrix} \mathbf{P} & \mathbf{Q} \\ \mathbf{Q}^t & \mathbf{R} \end{bmatrix},$$
where $\mathbf{P}$ *and* $\mathbf{R}$ *are square symmetric matrices. Then*
$$\lambda_{\max}(\mathbf{X}) + \lambda_{\min}(\mathbf{X}) \leq \lambda_{\max}(\mathbf{P}) + \lambda_{\max}(\mathbf{R}).$$

Proof   Let $\lambda = \lambda_{\min}(\mathbf{X})$ and take an arbitrary $\epsilon > 0$. Then $\mathbf{X}^* = \mathbf{X} - (\lambda - \epsilon)\mathbf{I}$ is a positive-definite symmetric matrix, partitioned in the same way as $\mathbf{X}$, with
$$\mathbf{P}^* = \mathbf{P} - (\lambda - \epsilon)\mathbf{I}, \quad \mathbf{Q}^* = \mathbf{Q}, \quad \mathbf{R}^* = \mathbf{R} - (\lambda - \epsilon)\mathbf{I}.$$
By applying the method of Rayleigh quotients to the matrix $\mathbf{X}^*$, it can be shown that
$$\lambda_{\max}(\mathbf{X}^*) \leq \lambda_{\max}(\mathbf{P}^*) + \lambda_{\max}(\mathbf{R}^*).$$
(See for instance *Linear Transformations* by H. L. Hamburger and M. E. Grimshaw (Cambridge, 1956), p. 77.) Thus, in terms of $\mathbf{X}$, $\mathbf{P}$ and $\mathbf{R}$, we have
$$\lambda_{\max}(\mathbf{X}) - (\lambda - \epsilon) \leq \lambda_{\max}(\mathbf{P}) - (\lambda - \epsilon) + \lambda_{\max}(\mathbf{R}) - (\lambda - \epsilon),$$
and since ϵ is arbitrary and $\lambda = \lambda_{\min}(\mathbf{X})$ we have the result. $\square$

Corollary 8.7   *Let* $\mathbf{A}$ *be a real symmetric matrix, partitioned into* t^2 *submatrices* $\mathbf{A}_{ij}$ *in such a way that the row and column partitions are the same; in other words, each diagonal sub-matrix* $\mathbf{A}_{ii}$ $(1 \leq i \leq t)$ *is square. Then*
$$\lambda_{\max}(\mathbf{A}) + (t-1)\lambda_{\min}(\mathbf{A}) \leq \sum_{i=1}^{t} \lambda_{\max}(\mathbf{A}_{ii}).$$

Proof   We prove this result by induction on t. It is true when $t = 2$, by the lemma. Suppose that it is true when $t = T - 1$; then we shall show that it holds when $t = T$. Let $\mathbf{A}$ be partitioned into T^2 submatrices, in the manner stated, and let $\mathbf{B}$ be the matrix $\mathbf{A}$ with the last row and column of submatrices deleted. By the lemma,
$$\lambda_{\max}(\mathbf{A}) + \lambda_{\min}(\mathbf{A}) \leq \lambda_{\max}(\mathbf{B}) + \lambda_{\max}(\mathbf{A}_{TT}),$$
and by the induction hypothesis,
$$\lambda_{\max}(\mathbf{B}) + (T-2)\lambda_{\min}(\mathbf{B}) \leq \sum_{i=1}^{T-1} \lambda_{\max}(\mathbf{A}_{ii}).$$
Now $\lambda_{\min}(\mathbf{B}) \geq \lambda_{\min}(\mathbf{A})$, as in the proof of Proposition 8.3. Thus, adding the two inequalities, we have the result for $t = T$, and the general result follows by induction. $\square$

We can now establish a lower bound for the chromatic number.

Theorem 8.8 (Hoffman 1970)   *For any graph* Γ, *whose edge-set is non-empty,*

$$\nu(\Gamma) \geq 1 + \frac{\lambda_{\max}(\Gamma)}{-\lambda_{\min}(\Gamma)}.$$

Proof   The vertex-set $V\Gamma$ can be partitioned into $\nu = \nu(\Gamma)$ colour-classes; consequently the adjacency matrix $\mathbf{A}$ of Γ can be partitioned into ν^2 submatrices, as in the preceding corollary. In this case, the diagonal submatrices $\mathbf{A}_{ii}$ $(1 \leq i \leq \nu)$ consist entirely of zeros, and so $\lambda_{\max}(\mathbf{A}_{ii}) = 0$ $(1 \leq i \leq \nu)$. Applying Corollary 8.7 we have

$$\lambda_{\max}(\mathbf{A}) + (\nu - 1)\lambda_{\min}(\mathbf{A}) \leq 0.$$

But, if Γ has at least one edge, then $\lambda_{\min}(\mathbf{A}) = \lambda_{\min}(\Gamma) < 0$. The result now follows. $\qquad\qquad\square$

In cases where the spectrum of a graph is known, Hoffman's bound can be very useful. Consider, for example, the graph Σ which arises from the classical configuration of 27 lines on a general cubic surface, in which each line meets 10 other lines. The vertices of Σ represent lines, and adjacent vertices represent skew lines, so that Σ is a regular graph with 27 vertices and degree 16. This is the graph with least eigenvalue -2 mentioned in **3g**. Since $\lambda_{\max}(\Sigma) = 16$ and $\lambda_{\min}(\Sigma) = -2$, Hoffman's bound is $\nu(\Sigma) \geq 1 + 16/2 = 9$, a result which would be difficult to establish by direct means. On the other hand, it is fairly easy to find a vertex-colouring using 9 colours (Haemers 1979), so Hoffman's bound leads to the exact answer $\nu(\Sigma) = 9$ in this case.

Additional Results

8a *The eigenvalues of a planar graph*   Let Γ be a planar connected graph. Then it follows from Theorem 8.8 and the four-colour theorem that

$$\lambda_{\min}(\Gamma) \leq -\frac{1}{3}\lambda_{\max}(\Gamma).$$

8b *Another bound for the chromatic number*   Let Γ be a regular graph of degree k with n vertices. In any colour-partition of Γ each colour-class has at most $n - k$ vertices; consequently $\nu(\Gamma) \geq n/(n-k)$. Cvetković (1972, see also **8h**) proved a corresponding result for any, not necessarily regular, graph:

$$\nu(\Gamma) \geq \frac{n}{n - \lambda_{\max}}.$$

8c *The second eigenvalue of the Laplacian*   The eigenvalues of a real symmetric matrix may be characterised in terms of the Rayleigh quotient. In particular, for the first non-zero eigenvalue μ_1 of the Laplacian matrix $\mathbf{Q}$, the characterisation asserts that

$$\mu_1 = \min_{\mathbf{u}^t \mathbf{x} = 0} R(\mathbf{Q}; \mathbf{x}),$$

where $\mathbf{u}$ is the all-1 vector, the eigenvector corresponding to μ_0. This provides a powerful method for finding upper bounds for μ_1. If we think of $\mathbf{x}$ as a real-valued function ξ defined on the vertex-set, the condition $\mathbf{u}^t \mathbf{x} = 0$ becomes $\sum \xi(v) = 0$, and for any function satisfying this condition we have (by a simple manipulation of the Rayleigh quotient)

$$\mu_1 \leq \sum_{\{v,w\} \in E} (\xi(v) - \xi(w))^2 \Big/ \sum_v \xi(v)^2.$$

8d *A spectral bound for the isoperimetric number*   Let δX be the cut defined by $X \subseteq V\Gamma$, and let $x = |X|$, $n = |V\Gamma|$. Define $\xi(v)$ to be $1/x$ if $v \in X$, and $-1/(n-x)$ otherwise. Then **8c** implies that

$$\mu_1 \leq |\delta X| \left(\frac{1}{x} + \frac{1}{n-x} \right).$$

It follows that, for the isoperimetric number defined in **4b**, we have $i(\Gamma) \geq \mu_1/2$, and in the k-regular case, $i(\Gamma) \geq (k - \lambda_1)/2$ (Alon and Milman 1985).

8e *Equipartitions*   Suppose that Γ is a graph with n vertices, and $\mathcal{X}$ is a partition of its vertex-set. Let $\delta\mathcal{X}$ denote the set of edges whose ends are in different parts. We say that $\mathcal{X}$ is an *equipartition* if each part has the same size p; then $n = pq$ where q is the number of parts. Biggs, Brightwell, and Tsoubelis (1992) showed that in this case

$$\mu_1 \leq |\delta\mathcal{X}| \left(\frac{2}{p(q-1)} \right).$$

8f *The odd graphs* (Biggs 1979)   Let k be a natural number greater than 1, and let S be a set of cardinality $2k-1$. The *odd graph* O_k is defined as follows: its vertices correspond to the subsets of S of cardinality $k-1$, and two vertices are adjacent if and only if the corresponding subsets are disjoint. (For example, $O_2 = K_3$, and O_3 is the Petersen graph.) O_k is a regular graph of degree k; when $k = 2$ its girth is 3, when $k = 3$ it is 5, and when $k \geq 4$ it is 6.

The spectrum of O_k can be obtained by the methods described in Chapters 20 and 21 (see **21b**). In particular, the largest eigenvalue $\lambda_{\max}$

is k, the next largest is $k - 2$, and the least is $1 - k$. Using Theorem 8.8 and **8d** respectively, we get the following lower bounds for the chromatic and isoperimetric numbers:

$$\nu(O_k) \geq 1 + \frac{k}{k-1} > 2; \quad i(O_k) \geq \frac{k - (k - 2)}{2} = 1.$$

To see how good these bounds are, let $V[\alpha, \beta]$ denote the set of vertices containing a given pair $\alpha, \beta \in S$; let $V[\alpha, \overline{\beta}]$ denote the set of vertices containing α but not β; and so on. Then the three sets $V[\overline{\alpha}, \beta]$, $V[\overline{\alpha}, \overline{\beta}]$, and $V[\alpha, \beta] \cup V[\alpha, \overline{\beta}]$ form a colour partition, so $\nu(O_k) = 3$. Furthermore, the cut defined by $X = V[\alpha, \beta] \cup V[\overline{\alpha}, \overline{\beta}]$ and its complement satisfies

$$\frac{|\delta X|}{|X|} = \frac{2 \binom{2k-3}{k-2}}{\binom{2k-3}{k-3} + \binom{2k-3}{k-1}} = \frac{k}{k-1}.$$

Thus $i(O_k) \leq 1 + (k - 1)^{-1}$. Further results about the odd graphs may be found in **17d**, **20b**, and **21b**.

8g *The Motzkin–Straus formula*   Consider the quadratic programming problem (QP)

$$\text{maximize} \quad \mathbf{x}^t \mathbf{A} \mathbf{x} \quad \text{subject to} \quad \mathbf{u}^t \mathbf{x} = 1, \ \mathbf{x} \geq \mathbf{0},$$

where $\mathbf{A}$ is the adjacency matrix of a graph Γ. Define the *support* of a feasible vector $\mathbf{x}$ to be the set of vertices v_i for which $x_i \neq 0$. It can be shown that, for an optimal $\mathbf{x}$ with minimal support, the support is a clique (a complete subgraph) in Γ. It follows that the maximum value for the QP is $1 - 1/\omega(\Gamma)$, where $\omega(\Gamma)$ is the size of the largest clique in Γ. This is the formula of Motzkin and Straus (1965). Putting $x_i = 1/n$ for $i = 1, 2, \ldots, n = |V\Gamma|$, and letting $m = |E\Gamma|$, we get

$$1 - \frac{1}{\omega(\Gamma)} \geq \frac{2m}{n^2}.$$

In particular, we have *Turán's Theorem*: if Γ has no triangles, then $m \leq n^2/4$.

8h *Another spectral bound*   Let s be the sum of the entries of the normalized eigenvector corresponding to $\lambda_{\max}$. Wilf (1985) observed that the Motzkin–Straus formula leads to the result $\omega \geq s^2/(s^2 - \lambda_{\max})$. Since $s^2 \leq n$ it follows that $\omega \geq n/(n - \lambda_{\max})$. Furthermore, the chromatic number ν cannot be less than ω, so this strengthens the result **8b** of Cvetković.

PART TWO

Colouring problems

9

The chromatic polynomial

Part Two is concerned with polynomial functions which represent certain numbers associated with graphs. The best-known example, the chromatic polynomial, is introduced in this chapter. It should be stressed that here we have to deal with general graphs, because some of the constructions fail when restricted to strict graphs.

Definition 9.1 Let Γ be a general graph with n vertices, and let u be a complex number. For each natural number r, let $m_r(\Gamma)$ denote the number of distinct colour-partitions of $V\Gamma$ into r colour-classes, and define $u_{(r)}$ to be the complex number $u(u-1)(u-2)\ldots(u-r+1)$. The *chromatic polynomial* of Γ is the function defined by

$$C(\Gamma; u) = \sum_{r=1}^{n} m_r(\Gamma) u_{(r)}.$$

Proposition 9.2 *If s is a natural number, then $C(\Gamma; s)$ is the number of vertex-colourings of Γ using at most s colours.*

Proof Every vertex-colouring of Γ in which exactly r colours are used gives rise to a colour-partition into r colour-classes. Conversely, for each colour-partition into r colours we can assign s colours to the colour-classes in $s(s-1)\ldots(s-r+1)$ ways. Hence the number of vertex-colourings in which s colours are available is $\sum m_r(\Gamma) s_{(r)} = C(\Gamma; s)$. $\square$

The simplest example is the chromatic polynomial of the complete graph K_n. Since every vertex of K_n is adjacent to every other one, the numbers of colour-partitions are

$$m_1(K_n) = m_2(K_n) = \ldots = m_{n-1}(K_n) = 0; \quad m_n(K_n) = 1.$$

Hence

$$C(K_n; u) = u(u-1)(u-2)\ldots(u-n+1).$$

Possibly the most important fact about the chromatic polynomial is that it is indeed a polynomial; in other words, the number of vertex-colourings of a graph, with a given number of colours available, is the value of a polynomial function. This is because the expressions $u_{(r)}$ which occur in the definition are themselves polynomials.

Some simple properties of the chromatic polynomial follow directly from its definition. For example, if Γ has n vertices, then $m_n(\Gamma) = 1$; hence $C(\Gamma; u)$ is a monic polynomial of degree n. Other results follow directly from Proposition 9.2 and the principle that a polynomial is uniquely determined by its values at an infinite set of natural numbers. For instance, if Γ is disconnected, with two components Γ_1 and Γ_2, then we can colour the vertices of Γ_1 and Γ_2 independently, and it follows that $C(\Gamma; s) = C(\Gamma_1; s)C(\Gamma_2; s)$ for any natural number s. Consequently

$$C(\Gamma; u) = C(\Gamma_1; u)C(\Gamma_2; u),$$

as elements of the ring of polynomials with integer coefficients.

Since u is a factor of $u_{(r)}$ for all $r \geq 1$, it follows that $C(\Gamma; 0) = 0$ for any general graph Γ. If Γ has c components, then the coefficients of $1 = u^0, u^1, \ldots, u^{c-1}$ are all zero, by virtue of the result on disconnected graphs in the previous paragraph. Also, if $E\Gamma \neq \emptyset$ then Γ has no vertex-colouring with just one colour, and so $C(\Gamma; 1) = 0$ and $u - 1$ is a factor of $C(\Gamma; u)$.

The problem of finding the chromatic number of a graph is part of the general problem of locating the zeros of its chromatic polynomial, because the chromatic number $\nu(\Gamma)$ is the smallest natural number ν which is not a zero of $C(\Gamma; u)$. This fact has stimulated some interesting work (see **9i**, **9j** and **9k** for example), but, as yet, elementary methods have proved more useful in answering questions about chromatic numbers.

The simplest method of calculating chromatic polynomials is a recursive technique. Suppose that Γ is a general graph and that e is an edge of Γ which is not a loop. The graph $\Gamma^{(e)}$ whose edge-set is $E\Gamma \setminus \{e\}$ and whose vertex-set is $V\Gamma$ is said to be obtained by *deleting* e; while the graph $\Gamma_{(e)}$ constructed from $\Gamma^{(e)}$ by identifying the two vertices incident with e in Γ, is said to be obtained by *contracting* e. We note that $\Gamma^{(e)}$

has one edge fewer than Γ, and $\Gamma_{(e)}$ has one edge and one vertex fewer than Γ, and so the following Proposition provides a method for calculating the chromatic polynomial by repeated reduction to 'smaller' graphs. This is known as the *deletion–contraction* method.

Proposition 9.3 *The chromatic polynomial satisfies the relation*
$$C(\Gamma; u) = C(\Gamma^{(e)}; u) - C(\Gamma_{(e)}; u).$$

Proof Consider the vertex-colourings of $\Gamma^{(e)}$ with s colours available. These colourings fall into two disjoint sets: those in which the ends of e are coloured differently, and those in which the ends of e are coloured alike. The first set is in bijective correspondence with the colourings of Γ, and the second set is in bijective correspondence with the colourings of $\Gamma_{(e)}$. Hence $C(\Gamma^{(e)}; s) = C(\Gamma; s) + C(\Gamma_{(e)}; s)$ for each natural number s, and the result follows. $\qquad\qquad\square$

Corollary 9.4 *If T is a tree with n vertices then*
$$C(T; u) = u(u - 1)^{n-1}.$$

Proof We prove this by induction, using the elementary fact that any tree with $n \geq 2$ vertices has a vertex (in fact, at least two vertices) of degree 1. The result is clearly true when $n = 1$. Suppose it is true when $n = N - 1$, and let T be a tree with N vertices, e an edge of T incident with a vertex of degree 1. Then $T^{(e)}$ has two components: an isolated vertex, and a tree with $N - 1$ vertices, the latter being $T_{(e)}$. Hence
$$C(T^{(e)}; u) = uC(T_{(e)}; u),$$
and using Proposition 9.3 and the induction hypothesis
$$C(T; u) = (u - 1)C(T_{(e)}; u) = (u - 1)u(u - 1)^{N-2} = u(u - 1)^{N-1}.$$
Hence the result is true when $n = N$, and for all n by the principle of induction. $\qquad\qquad\square$

The deletion–contraction method also yields the chromatic polynomial of a cycle graph C_n. If $n \geq 3$ the deletion of any edge from C_n results in a path graph P_n, which is a tree with n vertices, and the contraction of any edge results in a cycle graph C_{n-1}. Hence
$$C(C_n; u) = u(u - 1)^{n-1} - C(C_{n-1}; u).$$
Since $C_3 = K_3$, we have
$$C(C_3; u) = u(u - 1)(u - 2) = (u - 1)^3 - (u - 1).$$
We can solve the recursion given above, with this 'initial condition', to obtain the formula
$$C(C_n; u) = (u - 1)^n + (-1)^n(u - 1).$$

We now describe two other useful techniques for calculating chromatic polynomials. The first is concerned with the 'join' operation for graphs. Suppose Γ_1 and Γ_2 are two graphs; then we define their *join* $\Gamma_1 + \Gamma_2$ to be the graph with vertex-set and edge-set given by

$$V(\Gamma_1 + \Gamma_2) = V\Gamma_1 \cup V\Gamma_2;$$

$$E(\Gamma_1 + \Gamma_2) = E\Gamma_1 \cup E\Gamma_2 \cup \{\{x, y\} \mid x \in V\Gamma_1, y \in V\Gamma_2\}.$$

In other words, $\Gamma_1 + \Gamma_2$ consists of copies of Γ_1 and Γ_2 with additional edges joining every vertex of Γ_1 to every vertex of Γ_2.

Proposition 9.5   *The numbers of colour-partitions of* $\Gamma = \Gamma_1 + \Gamma_2$ *are given by*

$$m_i(\Gamma) = \sum_{j+l=i} m_j(\Gamma_1) m_l(\Gamma_2).$$

Proof   Since every vertex of Γ_1 is adjacent (in Γ) to every vertex of Γ_2, any colour-class of vertices in Γ is either a colour-class in Γ_1 or a colour-class in Γ_2. Hence the result.   □

Corollary 9.6   *The chromatic polynomial of the join* $\Gamma_1 + \Gamma_2$ *is*

$$C(\Gamma_1 + \Gamma_2; u) = C(\Gamma_1; u) \circ C(\Gamma_2; u),$$

where the $\circ$ *operation on polynomials signifies that we write each polynomial in the form* $\sum m_i u_{(i)}$ *and multiply as if* $u_{(i)}$ *were the power* u^i.   □

For example, the complete bipartite graph $K_{3,3}$ is the join $N_3 + N_3$, where N_n is the graph with n vertices and no edges. From Corollary 9.6, we have

$$
\begin{aligned}
C(K_{3,3}; u) &= u^3 \circ u^3 \\
&= (u_{(3)} + 3u_{(2)} + u_{(1)}) \circ (u_{(3)} + 3u_{(2)} + u_{(1)}) \\
&= u_{(6)} + 6u_{(5)} + 11u_{(4)} + 6u_{(3)} + u_{(2)} \\
&= u^6 - 9u^5 + 36u^4 - 75u^3 + 78u^2 - 31u.
\end{aligned}
$$

The chromatic polynomials of all complete multipartite graphs can be found in this way.

Another application of the method yields the chromatic polynomials of the graphs $N_1 + \Gamma$ and $N_2 + \Gamma$, sometimes known as the *cone* and *suspension* of Γ, and denoted by $c\Gamma$ and $s\Gamma$ respectively.

Proposition 9.7 *The chromatic polynomials of a cone and a suspension are given by*

$$C(c\Gamma; u) = uC(\Gamma; u - 1),$$

$$C(s\Gamma; u) = u(u - 1)C(\Gamma; u - 2) + uC(\Gamma; u - 1).$$

Proof Let $C(\Gamma; u) = \sum m_i u_{(i)}$. Using Corollary 9.6 and the fact that $u_{(i+1)} = u(u - 1)_{(i)}$, we have

$$C(c\Gamma; u) = C(N_1 + \Gamma; u) = u \circ C(\Gamma; u) = u_{(1)} \circ \sum m_i u_{(i)}$$

$$= \sum m_i u_{(i+1)} = u \sum m_i (u - 1)_{(i)} = uC(\Gamma; u - 1).$$

The second part is proved in a similar way, using the identity $u^2 = u_{(2)} + u_{(1)}$. $\qquad\square$

Another useful technique for the calculation of chromatic polynomials applies to graphs of the kind described in the next definition.

Definition 9.8 The general graph Γ is *quasi-separable* if there is a subset K of $V\Gamma$ such that the induced subgraph $\langle K \rangle$ is a complete graph and the induced subgraph $\langle V\Gamma \setminus K \rangle$ is disconnected. Γ is *separable* if $|K| \leq 1$; in this case either $K = \emptyset$ so that Γ is in fact disconnected, or $|K| = 1$, in which case we say that the single vertex of K is a *cut-vertex*.

It follows that in a quasi-separable graph Γ we have $V\Gamma = V_1 \cup V_2$, where $\langle V_1 \cap V_2 \rangle$ is complete and there are no edges in Γ joining $V_1 \setminus (V_1 \cap V_2)$ to $V_2 \setminus (V_1 \cap V_2)$. We shall refer to the pair (V_1, V_2) as a *quasi-separation* of Γ, or simply a *separation* if $|V_1 \cap V_2| \leq 1$.

A graph which is quasi-separable but not separable is shown in Figure 3; the relevant quasi-separation is given by $V_1 = \{1, 2, 4\}, V_2 = \{2, 3, 4\}$.

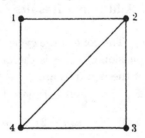

Figure 3: a quasi-separable graph.

Proposition 9.9 *If the graph Γ has quasi-separation (V_1, V_2) then*

$$C(\Gamma; u) = \frac{C(\langle V_1 \rangle; u) \, C(\langle V_2 \rangle; u)}{C(\langle V_1 \cap V_2 \rangle; u)}.$$

Proof If $V_1 \cap V_2$ is empty we make the convention that the denominator is 1, and the result is a consequence of the remark about disconnected graphs following Proposition 9.2. Suppose that $\langle V_1 \cap V_2 \rangle$ is a complete graph K_t, $t \geq 1$. Since Γ contains this complete graph, Γ has no vertex-colouring with fewer than t colours, and so $u_{(t)}$ is a factor of $C(\Gamma; u)$. For each natural number $s \geq t$, $C(\Gamma; s)/s_{(t)}$ is the number of ways of extending a given vertex-colouring of $\langle V_1 \cap V_2 \rangle$ to the whole of Γ, using at most s colours. Also, both $\langle V_1 \rangle$ and $\langle V_2 \rangle$ contain the complete graph $K_t = \langle V_1 \cap V_2 \rangle$, so $C(\langle V_i \rangle; s)/s_{(t)}$, $i \in \{1, 2\}$, has a corresponding interpretation. Since there are no edges in Γ joining $V_1 \setminus (V_1 \cap V_2)$ to $V_2 \setminus (V_1 \cap V_2)$, the extensions of a vertex-colouring of $\langle V_1 \cap V_2 \rangle$ to $\langle V_1 \rangle$ and to $\langle V_2 \rangle$ are independent. Hence

$$\frac{C(\Gamma; s)}{s_{(t)}} = \frac{C(\langle V_1 \rangle; s)}{s_{(t)}} \frac{C(\langle V_2 \rangle; s)}{s_{(t)}},$$

for all $s \geq t$. The corresponding identity for the polynomials follows. □

The formula of Proposition 9.9 is often useful in working out chromatic polynomials of small graphs. For instance, the graph shown in Figure 3 is 'two K_3's with a common K_2'; hence its chromatic polynomial is

$$\frac{u(u-1)(u-2)\, u(u-1)(u-2)}{u(u-1)} = u(u-1)(u-2)^2.$$

An important theoretical application of Proposition 9.9 will be described in Chapter 12.

Additional Results

9a *Wheels and pyramids* The cone of the cycle graph C_{n-1} is the *wheel* or *pyramid* W_n; the suspension of C_{n-2} is the *double pyramid* Π_n. The chromatic polynomials of these graphs are

$$C(W_n; u) = u(u-2)^{n-1} + (-1)^{n-1}u(u-2);$$

$$C(\Pi_n; u) = u(u-1)(u-3)^{n-2} + u(u-2)^{n-2} + (-1)^n u(u^2 - 3u + 1).$$

9b *The cocktail-party graphs* Let $p_s(u) = C(H_s; u)$, where H_s is the cocktail-party graph $K_{2,2,\dots,2}$ with $2s$ vertices. The polynomials $p_s(u)$ can be found from the recursion

$$p_1(u) = u^2, \quad p_s(u) = u(u-1)p_{s-1}(u-2) + up_{s-1}(u-1) \quad (s \geq 2).$$

9c *Ladders and Möbius ladders* The *ladder* L_h ($h \geq 3$) is a regular graph of degree 3 with $2h$ vertices $u_1, u_2, \ldots, u_h, v_1, v_2, \ldots, v_h$; the vertices $u_1, \ldots, u_h$ form a cycle of length h, as do the vertices $v_1, \ldots, v_h$, and the remaining edges are of the form $\{u_i, v_i\}$, $1 \leq i \leq h$. The Möbius ladders M_h were defined in **3e**. By systematic use of the deletion–contraction method Biggs, Damerell, and Sands (1972, see also **9i**) showed that

$$C(L_h; u) = (u^2 - 3u + 3)^h + (u - 1)\{(3 - u)^h + (1 - u)^h\} + u^2 - 3u + 1;$$

$$C(M_h; u) = (u^2 - 3u + 3)^h + (u - 1)\{(3 - u)^h - (1 - u)^h\} - 1.$$

9d *The chromatic polynomial characterizes trees* Corollary 9.4 implies that different graphs may have the same chromatic polynomial, since any two trees with the same number of vertices have this property. However, if Γ is a simple graph with n vertices, and $C(\Gamma; u) = u(u - 1)^{n-1}$, then Γ is a tree (Read 1968).

9e *Chromatically unique graphs* A graph is said to be *chromatically unique* if it is the only graph with its chromatic polynomial. From Corollary 9.4 we know that any tree with more than three vertices is not chromatically unique. Several families of graphs are known to be chromatically unique, among them the following.
 (a) The complete graphs, K_n.
 (b) The cycle graphs, C_n ($n \geq 3$).
 (c) The wheel graphs W_n for *odd* n.
It is known that W_6 and W_8 are not chromatically unique, but W_{10} is. See Li and Whitehead (1992) for this result and additional references.

9f *The chromatic polynomials of the regular polyhedra* The chromatic polynomials of the graphs formed by the vertices and edges of the five regular polyhedra in three dimensions are known. The graph of the tetrahedron is K_4, the graph of the octahedron is $H_3 = K_{2,2,2}$ (**9b**), and the graph of the cube is L_4 (**9c**). The chromatic polynomial of the icosahedron was computed by Whitney (1932b); after removing the factors $u(u - 1)(u - 2)(u - 3)$, it is

$$u^8 - 24u^7 + 260u^6 - 1670u^5 + 6999u^4 - 19698u^3 + 36408u^2 - 40240u + 20170.$$

The computation of the chromatic polynomial of the dodecahedron was first attempted by D.A. Sands (in an unpublished thesis, 1972) and Haggard (1976). In order to reduce the size of the coefficients it is convenient to express the result in the form

$$-u(u-1)(u-2)\sum_{i=0}^{17} c_i(1-u)^i.$$

In this form the coefficients c_i are all positive, and they are:

$$1,\ 10,\ 56,\ 230,\ 759,\ 2112,\ 5104,\ 10912,\ 20880,\ 35972,$$

$$55768,\ 77152,\ 93538,\ 96396,\ 80572,\ 50808,\ 21302,\ 4412.$$

9g *Interpolation formulae*   Suppose that two finite sequences of real numbers $m_0,\ m_1,\ldots,\ m_n$ and $p_0, p_1,\ldots, p_n$ are related by the rule

$$p_k = \sum_{r=0}^{k} m_r k_{(r)}.$$

Then there is an inverse formula giving the m's in terms of the p's, and this in turn leads to a formula for the polynomial $p(u)$ of degree n whose value at $k \in \{0, 1,\ldots, n\}$ is p_k:

$$m_r = \sum_{k=0}^{r} \frac{(-1)^{r-k}}{k!(r-k)!}p_k, \quad p(u) = \sum_{r=0}^{n} u_{(r)} \sum_{k=0}^{r} \frac{(-1)^{r-k}}{k!(r-k)!}p_k.$$

In particular we have formulae for the numbers of colour-partitions, and the chromatic polynomial, in terms of the numbers of k-colourings.

9h *Acyclic orientations* An orientation of a graph, as defined in Chapter 4, is said to be *acyclic* if it has no directed cycles. For example, on a tree with n vertices any orientation is acyclic, so there are 2^{n-1} acyclic orientations. Stanley (1973) showed that in general the number of acyclic orientations of Γ is the absolute value of $C(\Gamma; -1)$.

9i *Recursive families and chromatic roots*   As was remarked at the beginning of this chapter, the location of the zeros of a chromatic polynomial is a fundamental problem, because it subsumes the problem of finding the chromatic number. One of the few positive results in this direction is that the zeros for some families of graphs lie near certain curves in the complex plane. Biggs, Damerell and Sands (1972) defined a *recursive family* of graphs $\{\Gamma_n\}$ to be a sequence of graphs in which the polynomials $C(\Gamma_n; u)$ are related by a linear homogeneous recurrence, in which the coefficients are polynomials in u. In this case $C(\Gamma_n; u)$ can be expressed in the form

$$C(\Gamma_n; u) = \sum_{i=1}^{k} \alpha_i(u)[\lambda_i(u)]^n,$$

where the functions α_i and λ_i are not necessarily polynomials. For example, the ladders form a recursive family, and as in **9c** we have

$$\alpha_1(u) = 1, \ \alpha_2(u) = u - 1, \ \alpha_3(u) = u - 1, \ \alpha_4(u) = u^2 - 3u + 3;$$

$$\lambda_1(u) = u^2 - 3u + 3, \ \lambda_2(u) = 3 - u, \ \lambda_3(u) = 1 - u, \ \lambda_4(u) = 1.$$

Define a *chromatic root* of the family $\{\Gamma_n\}$ to be a complex number ζ for which there is an infinite sequence (u_n) such that u_n is a zero of $C(\Gamma_n; u)$ and $\lim u_n = \zeta$. Beraha, Kahane and Weiss (1980) obtained necessary and sufficient conditions for ζ to be a chromatic root, and Read (1990) explained how their results confirm empirical observations of Biggs, Damerell and Sands concerning the chromatic roots of the ladder graphs. It turns out that the chromatic roots of the ladders are 0, 1, together with the points lying on parts of two quartic curves and the line $\mathcal{R}u = 2$.

9j *Planar graphs* It is clear that the integers $0, 1, 2, 3$ are zeros of $C(\Gamma; u)$ for suitable planar graphs Γ, but the four-colour theorem tells us that the integer 4 is never a zero. The first result about non-integral zeros was obtained by Tutte (1970). He observed that there is often a zero close to $(3 + \sqrt{5})/2 = 2.6180\ldots$, and he proved that for any graph Γ with n vertices which triangulates the plane

$$C(\Gamma; (3 + \sqrt{5})/2) \leq \left(\frac{\sqrt{5} - 1}{2} \right)^{n-5}.$$

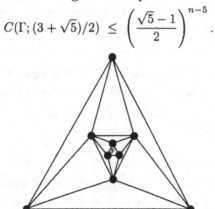

Figure 4: the iterated octahedron.

However this does not imply that $(3 + \sqrt{5})/2$ is a chromatic root of every family of plane triangulations. For example, the 'iterated octahedron' Δ_n (Figure 4) has chromatic polynomial

$$C(\Delta_n; u) = u(u - 1)(u - 2)(u^3 - 9u^2 + 29u - 32)^n.$$

Tutte's result applies to this family, and consequently all the graphs have a zero 'near' $2.6180\ldots$. But this zero is a constant, $2.5466\ldots$, independent of n; there is no zero which tends to $2.6180\ldots$ as $n \to \infty$.

9k *Chromatic roots of planar graphs*   Tutte's result (**9j**) led to speculation concerning the numbers $b_n = 2 + 2\cos(2\pi/n)$ as chromatic roots of families of planar graphs, based on the observations that $b_2 = 0$, $b_3 = 1$, $b_4 = 2$, $b_5 = 2.6180\ldots$, $b_6 = 3$, and $b_n \to 4$. Beraha and Kahane (1979) proved that 4 is indeed a chromatic root of a family of planar graphs, and Beraha, Kahane and Weiss (1980) proved the the same thing for b_5, b_7, and b_{10}. Concerning the numbers b_n in general, Tutte (1984) observes that 'their significance is not yet properly understood'.

9l *Zero-free intervals*   If Γ is a graph which triangulates the plane, there are no non-integral zeros ζ of $C(\Gamma; u)$ for which $\zeta < 2.5466\ldots$, the zero of the octahedron mentioned in **9j** (Woodall 1992).

9m *Confluence of the deletion–contraction method*   In the method of deletion and contraction we are free to choose any edge at each step. The fact that the order of choosing edges does not affect the final result is obvious, given the concrete interpretation of the chromatic polynomial in terms of colourings. However, the deletion–contraction method may be applied formally, as a set of so-called *rewriting rules*, and then it is necessary to prove that there is a 'normal form', independent of the order in which the rules are applied. This follows from two general properties of the rewriting rules, known as 'well-foundedness' and 'local confluence' (Yetter 1990).

9n *The umbral chromatic polynomial*   If P is a partition of an n-set in which there are α_i parts of size i, then we define the formal expression

$$\tau^\phi(P) = \phi_1^{\alpha_2} \ldots \phi_{n-1}^{\alpha_n}.$$

Given a graph Γ, let

$$m_r^\phi(\Gamma) = \sum_P \tau^\phi(P),$$

where the sum is over all colour-partitions P of $V\Gamma$ with r parts. Clearly, putting $\phi_1 = \phi_2 = \ldots = \phi_{n-1} = 1$ we obtain the ordinary $m_r(\Gamma)$, as defined on p. 63.

Ray and Wright (1992) show that the corresponding generalization of the chromatic polynomial is obtained by replacing the expressions $u_{(r)}$ by what are known as the *conjugate Bell polynomials* $b_r^\phi(u)$. Thus they define the *umbral chromatic polynomial*

$$C^\phi(\Gamma; u) = \sum_{r=1}^n m_r^\phi(\Gamma) b_r^\phi(u).$$

They obtain interpolation formulae like those in **9g** and analogues of other properties of the ordinary chromatic polynomial.

10

Subgraph expansions

It is clear that calculating the chromatic polynomial of a graph is at least as hard as finding its chromatic number. The latter problem is known to be difficult, in a technical sense which appears to correspond with practical experience. (More details may be found in the Additional Results at the end of Chapter 13.) There are nevertheless good reasons, both theoretical and practical, for studying methods of calculating the chromatic polynomial which are more sophisticated than those discussed in the previous chapter. These methods are based on the idea of an 'expansion' in terms of certain subgraphs.

Definition 10.1 The *rank polynomial* of a general graph Γ is the function defined by

$$R(\Gamma; x, y) = \sum_{S \subseteq E\Gamma} x^{r\langle S\rangle} y^{s\langle S\rangle},$$

where $r\langle S\rangle$ and $s\langle S\rangle$ are the rank and co-rank of the subgraph $\langle S\rangle$ of Γ. If we write $R(\Gamma; x, y) = \sum \rho_{rs} x^r y^s$ then ρ_{rs} is the number of subgraphs of Γ with rank r and co-rank s, and we say that the matrix (ρ_{rs}) is the *rank matrix* of Γ.

For example, the rank matrix of the graph $K_{3,3}$ is

$$\begin{bmatrix} 1 & & & & \\ 9 & & & & \\ 36 & & & & \\ 84 & 9 & & & \\ 117 & 45 & 6 & & \\ 81 & 78 & 36 & 9 & 1 \end{bmatrix}.$$

Here the rows are labelled by the values of the rank r from 0 to 5, and the columns are labelled by the values of the co-rank s from 0 to 4. We notice that since $r\langle S\rangle + s\langle S\rangle = |S|$ for all $S \subseteq E\Gamma$, an antidiagonal (sloping from bottom left to top right) corresponds to subgraphs with a fixed number t of edges, and consequently sums to the binomial coefficient $\binom{9}{t}$. We observe also that the number in the bottom left-hand corner (generally $\rho_{n-1,0}$, where n is the number of vertices) is just the tree-number of the graph. These facts mean that in this case very few entries need to be calculated explicitly.

As we shall see, several interesting functions can be obtained by assigning particular values to the indeterminates x and y in the rank polynomial. Trivially, putting $y = x$ gives $R(\Gamma; x, x) = (x+1)^{|E\Gamma|}$, because $r\langle S\rangle + s\langle S\rangle = |S|$ for all $S \subseteq E\Gamma$. The main result to be proved in this chapter is that by assigning certain values to x and y we obtain the chromatic polynomial.

For any natural number u, let $[u]$ denote the set $\{1, 2, \dots, u\}$, which we shall think of as a set of u colours, and let $[u]^X$ denote the set of all functions $\omega : X \to [u]$. For a general graph Γ, the set $[u]^{V\Gamma}$ contains some functions which are vertex-colourings of Γ with u colours available, and some functions which are not vertex-colourings since they violate the condition that adjacent vertices must receive different colours. In order to pick out the vertex-colourings we make the following definition.

Definition 10.2   For each $\omega \in [u]^{V\Gamma}$ we define the *indicator function* $\widehat{\omega} : E\Gamma \to \{0, 1\}$ as follows.

$$\widehat{\omega}(e) = \begin{cases} 1, & \text{if } e \text{ has vertices } v_1, v_2 \text{ such that } \omega(v_1) \neq \omega(v_2); \\ 0, & \text{otherwise.} \end{cases}$$

In particular, $\widehat{\omega}(e) = 0$ if e is a loop.

Lemma 10.3   *If Γ is a general graph and u is a natural number, then*

$$C(\Gamma; u) = \sum_{\omega \in [u]^{V\Gamma}} \prod_{e \in E\Gamma} \widehat{\omega}(e).$$

Proof   The product $\prod \widehat{\omega}(e)$ is zero unless $\widehat{\omega}(e) = 1$ for all $e \in E\Gamma$, and this is so only if ω is a vertex-colouring of Γ. Thus the sum of these products is the number of vertex-colourings of Γ using at most u colours. The result follows from Proposition 9.2. $\square$

Theorem 10.4   *The chromatic polynomial of a graph Γ with n vertices has an expansion in terms of subgraphs as follows:*

$$C(\Gamma; u) = \sum_{S \subseteq E\Gamma} (-1)^{|S|} u^{n-r\langle S\rangle}.$$

Proof For any natural number u we have

$$C(\Gamma; u) = \sum_{\omega \in [u]^{V\Gamma}} \prod_{e \in E\Gamma} \widehat{\omega}(e) = \sum_{\omega \in [u]^{V\Gamma}} \prod_{e \in E\Gamma} \{1 + (\widehat{\omega}(e) - 1)\}.$$

Expanding the product of terms $1 + f(e)$ we obtain a sum of $2^{|E\Gamma|}$ expressions $\prod f(e)$, one for each subset $S \subseteq E\Gamma$. That is,

$$C(\Gamma; u) = \sum_{\omega \in [u]^{V\Gamma}} \sum_{S \subseteq E\Gamma} \prod_{e \in S} (\widehat{\omega}(e) - 1).$$

We now switch the order in the double sum. For each $S \subseteq E\Gamma$ let $VS = V\langle S \rangle$; then any function from VS to $[u]$ is the restriction to VS of $u^{|V\Gamma \setminus VS|}$ functions from $V\Gamma$ to $[u]$. Thus

$$\sum_{\omega \in [u]^{V\Gamma}} \sum_{S \subseteq E\Gamma} \prod_{e \in S} (\widehat{\omega}(e) - 1) = \sum_{S \subseteq E\Gamma} u^{|V\Gamma \setminus VS|} \sum_{\omega \in [u]^{VS}} \prod_{e \in S} (\widehat{\omega}(e) - 1).$$

Consider the product $\prod(\widehat{\omega}(e) - 1)$ over all edges $e \in S$. If the product is non-zero, $\widehat{\omega}(e)$ must be 0 for each $e \in S$, which means that ω is constant on each component of $\langle S \rangle$. In this case the value of the product is $(-1)^{|S|}$. If $\langle S \rangle$ has c components there are u^c such functions ω; hence the sum of the product over all $u^{|VS|}$ functions $\omega : VS \to [u]$ is $(-1)^{|S|} u^c$. The result follows from the equation

$$|V\Gamma \setminus VS| + c = n - |VS| + c = n - r\langle S \rangle.$$

$\square$

Corollary 10.5 *The chromatic polynomial and the rank polynomial of a general graph Γ with n vertices are related by the identity*

$$C(\Gamma; u) = u^n R(\Gamma; -u^{-1}, -1).$$

If the chromatic polynomial is

$$C(\Gamma; u) = b_0 u^n + b_1 u^{n-1} + \ldots + b_{n-1} u + b_n,$$

then the coefficients b_i can be expressed in terms of the entries in the rank matrix, as follows:

$$(-1)^i b_i = \sum_j (-1)^j \rho_{ij}.$$

Proof The identity between the polynomials follows directly from Theorem 10.4 and the definition of the rank polynomial. In terms of the coefficients, we have

$$\sum_i b_i u^{n-i} = C(\Gamma; u) = u^n R(\Gamma; -u^{-1}, -1)$$

$$= u^n \sum_{r,s} \rho_{rs} (-u)^{-r} (-1)^s$$

$$= \sum_r \sum_s (-1)^{r+s} \rho_{rs} u^{n-r}.$$

Equating coefficients of powers of u, and rearranging the signs, we have the result stated above.   □

The formula for the coefficients expresses b_i as an alternating sum of the entries in the ith row of the rank matrix. This formula was first studied by Birkhoff (1912), in the original paper on chromatic polynomials, and Whitney (1932a). For example, from the rank matrix for $K_{3,3}$ given above we have

$$b_1 = -9, \quad b_2 = 36, \quad b_3 = -84 + 9 = -75, \quad b_4 = 117 - 45 + 6 = 78,$$

$$b_5 = -81 + 78 - 36 + 9 - 1 = -31.$$

This checks with the result obtained in Chapter 9 by a different method:

$$C(K_{3,3}; u) = u^6 - 9u^5 + 36u^4 - 75u^3 + 78u^2 - 31u.$$

Proposition 10.6 *Let Γ be a strict graph of girth g, having m edges and η cycles of length g. Then, with the above notation for the coefficients of the chromatic polynomial of Γ, we have*

(1) $(-1)^i b_i = \binom{m}{i}$ *for* $i = 0, 1, \ldots, g-2$;
(2) $(-1)^{g-1} b_{g-1} = \binom{m}{g-1} - \eta$.

Proof A subgraph of Γ with rank $i \leq g - 2$ must have co-rank zero, since Γ has no cycles with fewer than g edges. Thus, for all $i \leq g - 2$, we have $\rho_{i0} = \binom{m}{i}$ and $\rho_{ij} = 0$ if $j > 0$. Further, the only subgraphs of Γ with rank $g - 1$ are the $\binom{m}{g-1}$ forests with $g - 1$ edges (which have co-rank zero), and the η cycles with g edges (which have co-rank 1). Thus,

$$\rho_{g-1,0} = \binom{m}{i}, \quad \rho_{g-1,1} = \eta, \quad \rho_{g-1,j} = 0, \quad \text{if} \quad j > 1.$$

The result follows from the expression for the coefficients of the chromatic polynomial.   □

We observe that for a strict graph the girth g is at least 3, so the coefficient of u^{n-1} in the chromatic polynomial is $-m$, where n and m are the numbers of vertices and edges respectively.

The formula for the coefficients of the chromatic polynomial is an alternating sum, and its use involves counting many subgraphs which 'cancel out' in the final result. Whitney (1932a) discovered a reduction which involves counting fewer subgraphs. His result also shows that the non-zero coefficients of the chromatic polynomial alternate in sign; that is, $(-1)^i b_i$ is always positive. Let Γ be a simple graph whose edge-set $E\Gamma = \{e_1, e_2, \ldots, e_m\}$ is ordered by the natural order of subscripts. This ordering is to remain fixed throughout our discussion. A *broken cycle* in Γ is the result of removing the first edge from some cycle; in other words, it is a subset B of $E\Gamma$ such that for some edge e_l we have

(1) $B \cup \{e_l\}$ is a cycle in Γ;   (2) $i > l$ for each edge $e_i \in B$.

The next proposition expresses the coefficients of the chromatic polynomial in terms of the subgraphs which contain no broken cycles; clearly such subgraphs contain no cycles, and so they are forests.

Proposition 10.7 (Whitney 1932a)   *Let Γ be a strict graph whose edge-set is ordered as above, and let $C(\Gamma; u) = \sum b_i u^{n-i}$. Then $(-1)^i b_i$ is the number of subgraphs of Γ which have i edges and contain no broken cycles.*

Proof   Suppose $B_1, B_2, \ldots, B_t$ is a list of the broken cycles of Γ, in dictionary order based on the ordering of $E\Gamma$. Let f_i ($1 \leq i \leq t$) denote the edge which, when added to B_i, completes a cycle. The edges f_i are not necessarily all different, but, because of the way in which the broken cycles are ordered, it follows that f_i is not in B_j when $j \geq i$.

Define Σ_0 to be the set of subgraphs of Γ containing no broken cycle, and for $1 \leq h \leq t$ define Σ_h to be the set of subgraphs containing B_h but not $B_{h+1}, B_{h+2}, \ldots, B_t$. Then $\Sigma_0, \Sigma_1, \ldots, \Sigma_t$ is a partition of the set of all subgraphs of Γ. We claim that, in the expression

$$(-1)^i b_i = \sum (-1)^j \rho_{ij},$$

the total contribution to the sum from $\Sigma_1, \ldots, \Sigma_t$ is zero.

Suppose S is a subset of $E\Gamma$ not containing f_h; then S contains B_h if and only if $S \cup \{f_h\}$ contains B_h. Further, S contains B_i ($i > h$) if and only if $S \cup \{f_h\}$ contains B_i, since f_h is not in B_i. Thus if one of the subgraphs $\langle S \rangle, \langle S \cup \{f_h\} \rangle$ is in Σ_h, then both are in Σ_h. They have the same rank, but their co-ranks differ by one, and so their contributions to the alternating sum cancel. Consequently, we need only consider the contribution of Σ_0 to $\sum (-1)^j \rho_{ij}$. Since a subgraph $\langle S \rangle$ in Σ_0 is a forest, it has co-rank $j = 0$ and rank $i = |S|$, whence the result.   □

Corollary 10.8   *Let Γ be a strict graph with rank r. Then the co-*

efficients of $C(\Gamma; u)$ alternate strictly in sign; that is, $(-1)^i b_i > 0$ for $i = 0, 1 \ldots, r$.

Proof The characterization of Proposition 10.7 shows that $(-1)^i b_i \geq 0$ for $0 \leq i \leq n$. In order to obtain the strict inequality we must show that there is a subgraph with i edges and containing no broken cycle, for $i = 0, 1, \ldots, r$. Suppose we successively remove edges from Γ in such a way that at least one cycle is destroyed at each stage; this process stops when we reach a subgraph $\langle F \rangle$ of Γ with $|F| = r$ and $s\langle F \rangle = 0$. Let us order the edges of Γ so that the edges in F come first. Then $\langle F \rangle$ contains no broken cycle, and any subset of F generates a subgraph containing no broken cycle. Thus we have produced the required subgraphs, and the result follows. □

Recall that, at the beginning of Chapter 9, we observed that $b_i = 0$ if $i = n, n-1, \ldots, n-(c-1)$, where $n = |V\Gamma|$ and Γ has c components. That is, $b_i = 0$ if $i = r + 1, \ldots, n$. Thus we have shown that the coefficients of the chromatic polynomial alternate strictly, and then become zero.

Additional Results

10a *Inequalities for the coefficients of the chromatic polynomial* If Γ is a connected strict graph with n vertices and m edges, and $C(\Gamma; u) = \Sigma b_i u^{n-i}$, then

$$\binom{n-1}{i} \leq (-1)^i b_i \leq \binom{m}{i}.$$

10b *Codichromatic graphs* An example of two non-isomorphic general graphs having the same rank matrix was found in the 1930's by Marion C. Gray (see Figure 5).

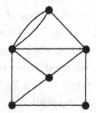 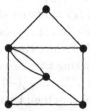

Figure 5: two general graphs with the same rank matrix.

Tutte (1974) drew attention to this work and constructed pairs of strict graphs which have the same rank matrix.

10c *V-functions* A function f defined on isomorphism classes of graphs and taking values in a ring A is a *V-function* if it satisfies the following conditions.
(a) If Γ is empty then $f(\Gamma) = 1$.
(b) If Γ is the union of disjoint graphs Γ_1, Γ_2, then $f(\Gamma) = f(\Gamma_1)f(\Gamma_2)$.
(c) If e is any edge of Γ which is not a loop, then f satisfies the deletion-contraction formula
$$f(\Gamma) = f(\Gamma^{(e)}) + f(\Gamma_{(e)}).$$

It is easy to see that the chromatic polynomial and the rank polynomial, suitably normalized, are V-functions. The most general V-function is constructed as follows. For any sequence $\mathbf{i} = i_0, i_1, i_2, \ldots$ of non-negative integers, with finite sum, let $v(\Gamma, \mathbf{i})$ be the number of spanning subgraphs of Γ which have i_k components of co-rank k, for $k \geq 0$. Let $\mathbf{s} = (s_0, s_1, s_2, \ldots)$ be any infinite sequence of elements of A and let
$$f_{\mathbf{s}}(\Gamma) = \sum_{\mathbf{i}} v(\Gamma, \mathbf{i}) \prod_{k=0}^{\infty} s_k^{i_k}.$$
Then $f_{\mathbf{s}}$ is a V-function, and every V-function can be written in this way (Tutte 1947b).

10d *The rank polynomial as a V-function* By taking the ring A to be the ring of polynomials with integer coefficients in two indeterminates x, y and $\mathbf{s}$ to be the sequence defined by $s_i = xy^i$, we obtain the rank polynomial (with suitable normalization) as a V-function.

10e *Homeomorphic graphs* The operation of replacing an edge with ends u, v by two edges with ends u, w and w, v, where w is a new vertex, is known as *subdividing* the edge. Regarding a graph as a topological space in the obvious way, it is clear that subdividing an edge results in a graph homeomorphic to the original one. In general, two graphs are homeomorphic if they can both be obtained from the same graph by sequences of subdivisions. A graph function is said to be a *topological invariant* if its values on homeomorphic graphs are equal. It can be shown that a non-trivial V-function f is a topological invariant if and only if $f(K_1) = -1$.

10f *Interaction models* The formula obtained in Lemma 10.3 arises naturally in theoretical physics. The vertices of the graph $\Gamma = (V, E)$

are 'particles', each of which which can have one of u attributes, so that a 'state' of the system is a function $\omega : V \to [u]$. Each pair of adjacent vertices, corresponding to an edge $e \in E$, has an interaction $i_\omega(e)$ which depends on the state ω, and the 'weight' $I(\omega)$ is the product of the interactions $i_\omega(e)$. The *partition function* is the sum of all the weights:

$$Z(\Gamma) = \sum_\omega I(\omega) = \sum_{\omega \in [u]^V} \prod_{e \in E} i_\omega(e).$$

The chromatic polynomial is the special case arising when i_ω is the indicator function $\widehat{\omega}$ as in Definition 10.2: that is, $i_\omega(e)$ is 1 if the ends of e have different attributes, and is 0 otherwise. In general, the value of u and the function i_ω determine an *interaction model*. An account of the properties of such models was given by Biggs (1977b).

10g *The Ising and Potts models*   Using the 'interaction model' terminology, suppose $i_\omega(e)$ is α if the ends of e have the same attribute in the state ω, and 1 otherwise. For general u this is known as the *Potts model*, and the special case $u = 2$ is known as the *Ising model*. The partition function for the Potts model can be expanded as a rank function:

$$Z_{Potts}(\Gamma) = \sum_{S \subseteq E\Gamma} u^{n-r\langle S\rangle}(\alpha - 1)^{|S|} = u^n R(\Gamma; \frac{\alpha - 1}{u}, \alpha - 1).$$

10h *A general form of the subgraph expansion*   An interaction model for which $i_\omega(e)$ takes only two values, one when the ends of e have the same attribute and another when they have different attributes, is said to be a *resonant* model. The expansions in terms of the rank function described above can be generalized to any resonant model in the following way. Let F and G be resonant models for which the two (distinct) values of $i_\omega(e)$ are f_1, f_0 and g_1, g_0 respectively, and let $f_i = \gamma g_i + \delta$, $i = 0, 1$. Then $Z_F(\Gamma)$ can be expanded in terms of the values of Z_G on the subgraphs of Γ as follows:

$$\frac{Z_F(\Gamma)}{u^{|V\Gamma|}} = \delta^{|E\Gamma|} \sum_{S \subseteq E\Gamma} \left(\frac{\gamma}{\delta}\right)^{|S|} \frac{Z_G(\langle S\rangle)}{u^{|S|}}.$$

10i *Another expansion of the chromatic polynomial*   Nagle (1971) obtained the following expansion:

$$C(\Gamma; u) = \sum_{S \subseteq E\Gamma} u^{|V\Gamma \setminus VS|}(1 - u^{-1})^{|E\Gamma \setminus S|} N(\langle S\rangle; u),$$

where the function N has the following properties: (a) if Γ has an isthmus, then $N(\Gamma; u) = 0$; (b) N is a topological invariant.

11

The multiplicative expansion

In this chapter and the next one we shall investigate expansions of the chromatic polynomial which involve relatively few subgraphs in comparison with the expansion of Chapter 10. The idea first appeared in the work of Whitney (1932b), and it was developed independently by Tutte (1967) and researchers in theoretical physics who described the method as a 'linked-cluster expansion' (Baker 1971). The simple version given here is based on a paper by the present author (Biggs 1973a). There are other approaches which use more algebraic machinery; see Biggs (1978) and **11e**.

We begin with some definitions. Recall that if a connected graph Γ is separable then it has a certain number of cut-vertices, and the removal of any cut-vertex disconnects the graph. A non-separable subgraph of Γ which is non-empty and maximal (considered as a subset of the edges) is known as a *block*. Every edge is in just one block, and we may think of Γ as a set of blocks 'stuck together' at the cut-vertices. In the case of a disconnected graph we define the blocks to be the blocks of the components. It is worth remarking that this means that isolated vertices are disregarded, since every block must have at least one edge.

Let Y be a real-valued function defined for all graphs, and having the following two properties.

*P*1: $Y(\Gamma) = 1$ if Γ has no edges;
*P*2: $Y(\Gamma)$ is the product of the numbers $Y(B)$ taken over all blocks B of Γ.

Given such a Y, let X be the real-valued function defined by

$$X(\Gamma) = \sum_{S \subseteq E\Gamma} Y\langle S \rangle.$$

An example of a function satisfying $P1$ and $P2$ is obtained by taking $Y(\Gamma) = x^{r(\Gamma)} y^{s(\Gamma)}$, where x and y are a given pair of real numbers, in which case the corresponding X is (an evaluation of) the rank polynomial of Γ. The fact that this Y satisfies $P2$ is a consequence of the equations

$$r(\Gamma) = \sum r(B), \quad s(\Gamma) = \sum s(B),$$

where the sums are taken over the set of blocks B of Γ.

Lemma 11.1   If the function Y satisfies $P1$ and $P2$, then the corresponding function X satisfies the same properties.

Proof   ($P1$) If Γ has no edges then the sum occurring in the definition of X contains only one term, $Y\langle \emptyset \rangle$, which is 1.

($P2$) Suppose Γ has just two blocks Γ_1 and Γ_2, with edge-sets E_1 and E_2. Then for any $S \subseteq E\Gamma$, the sets $S_1 = S \cap E_1$ and $S_2 = S \cap E_2$, are such that $S = S_1 \cup S_2$ and $S_1 \cap S_2 = \emptyset$. Thus the blocks of $\langle S \rangle$ in Γ are the blocks of $\langle S_1 \rangle$, regarded as a subgraph of Γ_1, together with the blocks of $\langle S_2 \rangle$ regarded as a subset of Γ_2. By $P2$ we have

$$Y\langle S \rangle_\Gamma = Y\langle S_1 \rangle_{\Gamma_1} Y\langle S_2 \rangle_{\Gamma_2}.$$

(This equation remains true if either or both of S_1, S_2 are empty, by $P1$.) Consequently,

$$X(\Gamma) = \sum_{S \subseteq E\Gamma} Y\langle S \rangle_\Gamma = \sum_{S_1 \subseteq E_1} \sum_{S_2 \subseteq E_2} Y\langle S_1 \rangle_{\Gamma_1} Y\langle S_2 \rangle_{\Gamma_2}$$

$$= \sum_{S_1 \subseteq E_1} Y\langle S_1 \rangle_{\Gamma_1} \sum_{S_2 \subseteq E_2} Y\langle S_2 \rangle_{\Gamma_2} = X(\Gamma_1) X(\Gamma_2).$$

If Γ has $b > 2$ blocks we have a similar argument, taking Γ_1 to be the first $b - 1$ blocks. Hence the general result follows by induction.   □

We shall now transform the sum $X(\Gamma)$ into a product, using exponential and logarithmic functions. We require also the fundamental identity underlying the 'principle of inclusion and exclusion': that is

$$\sum_{I \subseteq J} (-1)^{|I|} = 0,$$

provided that J is not the empty set.

Definition 11.2   Let (X, Y) be a pair of functions as above, and suppose that the values of X are positive. Then the *logarithmic transform* of the pair (X, Y) is the pair of functions $(\widetilde{X}, \widetilde{Y})$ defined by

$$\widetilde{X}(\Gamma) = (-1)^{|E\Gamma|} \sum_{S \subseteq E\Gamma} (-1)^{|S|} \log X\langle S \rangle, \quad \widetilde{Y}(\Gamma) = \exp \widetilde{X}(\Gamma).$$

Proposition 11.3 *Let Γ be a general graph. If Γ has no edges, or if Γ is separable and has no isolated vertices, then $\widetilde{X}(\Gamma) = 0$.*

Proof If $E\Gamma$ is empty then $X(\Gamma) = 1$, and consequently $\widetilde{X}(\Gamma) = 0$. Suppose that Γ has no isolated vertices and is separable. Then either it is disconnected or it is connected and has at least one cut-vertex. In either case it can be expressed as the union of two subgraphs $\langle E_1 \rangle$ and $\langle E_2 \rangle$ with E_1 and E_2 non-empty and disjoint. For $S \subseteq E\Gamma$, we have

$$X\langle S \rangle = X\langle S_1 \rangle X\langle S_2 \rangle,$$

where $S_1 = S \cap E_1$ and $S_2 = S \cap E_2$, and so $\log X\langle S \rangle = \log X\langle S_1 \rangle + \log X\langle S_2 \rangle$. This justifies the following calculation.

$$\begin{aligned}
(-1)^{|E\Gamma|} \widetilde{X}(\Gamma) &= \sum_{S \subseteq E\Gamma} (-1)^{|S|} \log X\langle S \rangle \\
&= \sum_{S_1 \subseteq E_1} \sum_{S_2 \subseteq E_2} (-1)^{|S_1| + |S_2|} (\log X\langle S_1 \rangle + \log X\langle S_2 \rangle) \\
&= \left[\sum_{S_1 \subseteq E_1} (-1)^{|S_1|} \log X\langle S_1 \rangle \sum_{S_2 \subseteq E_2} (-1)^{|S_2|} \right] \\
&\quad + \left[\sum_{S_2 \subseteq E_2} (-1)^{|S_2|} \log X\langle S_2 \rangle \sum_{S_1 \subseteq E_1} (-1)^{|S_1|} \right].
\end{aligned}$$

Both E_1 and E_2 are non-empty, so the fundamental inclusion–exclusion identity stated above implies that the entire expression is zero, and we have the result. □

Theorem 11.4 *Let Γ be a non-separable graph, and let $(\widetilde{X}, \widetilde{Y})$ be the logarithmic transform of the pair (X, Y). Then $X(\Gamma)$ has a multiplicative expansion*

$$X(\Gamma) = \prod_{S \subseteq E\Gamma} \widetilde{Y}\langle S \rangle,$$

in which $\widetilde{Y}$ is equal to 1 (and so may be ignored) for separable subgraphs of Γ.

Proof The fact that $\widetilde{Y}\langle S \rangle = \exp \widetilde{X}\langle S \rangle = 1$ for separable subgraphs $\langle S \rangle$ follows from the previous lemma, since, by definition, a subgraph has no isolated vertices.

We shall prove that

$$\log X(\Gamma) = \sum_{S \subseteq E\Gamma} \tilde{X}\langle S \rangle,$$

from which the theorem follows by taking exponentials. Now, from the definition of $\tilde{X}$,

$$\sum_{S \subseteq E\Gamma} \tilde{X}\langle S \rangle = \sum_{S \subseteq E\Gamma} (-1)^{|S|} \sum_{R \subseteq S} (-1)^{|R|} \log X\langle R \rangle,$$

and $\langle R \rangle$ as a subgraph of $\langle S \rangle$ is identical with $\langle R \rangle$ as a subgraph of Γ. Writing $Y = S \setminus R$ the right-hand side becomes

$$\sum_{R \subseteq E\Gamma} \sum_{Y \subseteq E\Gamma \setminus R} (-1)^{|R|+|Y|} (-1)^{|R|} \log X\langle R \rangle$$

$$= \sum_{R \subseteq E\Gamma} \log X\langle R \rangle \sum_{Y \subseteq E\Gamma \setminus R} (-1)^{|Y|}.$$

The inner sum is non-zero only when $E\Gamma \setminus R = \emptyset$; that is, when $R = E\Gamma$. Thus the expression reduces to $\log X\langle E\Gamma \rangle = \log X(\Gamma)$, as required.

$\square$

We now apply the general theory of the logarithmic transform to the particular case of the chromatic polynomial. We take the function Y to be

$$Y_u(\Gamma) = (-1)^{|E\Gamma|} u^{-r(\Gamma)}.$$

This satisfies $P1$ and $P2$ and, by Theorem 10.4, the corresponding X function is

$$X_u(\Gamma) = u^{-|V\Gamma|} C(\Gamma; u).$$

Lemma 11.5 *Let (X_u, Y_u) denote the particular pair of functions given above. Then, for a given graph Γ, $\tilde{X}_u(\Gamma)$ and $\tilde{Y}_u(\Gamma)$ can be defined for all sufficiently large integers u.*

Proof In order to define $\tilde{X}_u(\Gamma)$ satisfactorily we must ensure that $\log X_u\langle S \rangle$ is defined for all subsets $S \subseteq E\Gamma$. Now if u is an integer greater than the chromatic number of Γ, it is clear that $C(\langle S \rangle; u)$ is positive and so the logarithm of $X_u\langle S \rangle = u^{-|V\langle S \rangle|} C(\langle S \rangle; u)$ is defined.

$\square$

We can eliminate the logarithmic and exponential functions from the general definition of $\tilde{Y}$, obtaining

$$\tilde{Y}(\Gamma) = \prod_{S \subseteq E\Gamma} \{X\langle S \rangle\}^{\epsilon(S)}, \quad \text{where} \quad \epsilon(S) = (-1)^{|E\Gamma \setminus S|}.$$

For the particular case $\widetilde{Y}_u$ we get

$$\widetilde{Y}_u(\Gamma) = \prod_{S \subseteq E\Gamma} \{u^{-|V\langle S\rangle|} C(\langle S\rangle; u)\}^{\epsilon(S)},$$

which is valid for all sufficiently large positive integers u. The product formula shows that $\widetilde{Y}_u$ is a rational function in its domain of definition.

We shall find it convenient to deal separately with the $\widetilde{Y}_u$ function for a single edge, that is $\widetilde{Y}_u(K_2)$. It is easy work this out explicitly: $\widetilde{Y}_u(K_2) = 1 - u^{-1}$.

Proposition 11.6 *For every non-separable graph Λ having more than one edge there is a rational function $q(\Lambda; u)$ such that the chromatic polynomial of a graph Γ has a multiplicative expansion*

$$C(\Gamma; u) = u^{|V\Gamma|}(1 - u^{-1})^{|E\Gamma|} \prod q(\Lambda; u),$$

where the product is taken over all those non-separable subgraphs Λ of Γ which have more than one edge.

Proof We have seen that, if (X_u, Y_u) is the pair defined by

$$Y_u(\Gamma) = (-1)^{|E\Gamma|} u^{-r(\Gamma)}, \quad X_u(\Gamma) = u^{-|V\Gamma|} C(\Gamma; u),$$

then $\widetilde{Y}_u$ is defined for all subgraphs of Γ provided u is a sufficiently large integer, and $\widetilde{Y}_u(K_2) = (1 - u^{-1})$. Setting $q(\Lambda; u) = \widetilde{Y}_u(\Lambda)$ when $|E\Lambda| > 1$ and applying Theorem 11.4 we see that the identity holds for an infinite set of values of u. Since both sides are holomorphic functions, they are identical. $\square$

The functions $q(\Gamma; u)$ can be found explicitly for certain standard graphs. For example, for the cycle graph C_n the only subgraph occuring in the product is C_n itself; hence

$$C(C_n; u) = u^n (1 - u^{-1})^n q(C_n; u).$$

By a result of Chapter 9, the left-hand side is $(u - 1)^n + (-1)^n(u - 1)$, so that

$$q(C_n; u) = 1 + \frac{(-1)^n}{(u - 1)^{n-1}}.$$

This simple calculation highlights an apparent circularity which arises if we propose to use the multiplicative expansion to calculate chromatic polynomials. The difficulty is that the right-hand side of the multiplicative expansion of $C(\Gamma; u)$ contains a term $q(\Gamma; u)$, and we have, as yet, no way of finding $q(\Gamma; u)$ without prior knowledge of $C(\Gamma; u)$. In the next chapter it will be shown that this seemingly fundamental objection can be surmounted by means of a few simple observations. We shall also obtain a version of Proposition 11.6 in which the number of subgraphs involved is reduced still further.

Additional Results

11a *The q function of a crossed cycle*   Let C_n^+ denote a graph constructed from the cycle graph C_n by the addition of one edge joining two distinct vertices which are not adjacent in C_n. Then

$$q(C_n^+; u) = \{q(C_n; u)\}^{-1} = 1 + \frac{(-1)^{n-1}}{(u-1)^{n-1} + (-1)^n}.$$

11b *Theta graphs* (Baker 1971)   Let $\Theta_{r,s,t}$ denote the graph consisting of two vertices joined by three disjoint paths of length r, s, and t. $\Theta_{r,s,t}$ has $n = r + s + t - 1$ vertices and $r + s + t$ edges, and $q(\Theta_{r,s,t}; u)$ is

$$\frac{1 - (1-u)^{r-n} - (1-u)^{s-n} - (1-u)^{t-n} + (2-u)(1-u)^{-n}}{(1 - (1-u)^{r-n})(1 - (1-u)^{s-n})(1 - (1-u)^{t-n})}.$$

11c *The multiplicative expansion of the rank polynomial*   If $Y(\Gamma) = x^{r(\Gamma)} y^{s(\Gamma)}$ then $X(\Gamma) = R(\Gamma; x, y)$, and the logarithmic transform applied to the pair (X, Y) leads to a multiplicative expansion

$$R(\Gamma; x, y) = (1 + x)^{|E\Gamma|} \prod \widetilde{Y}(\Lambda; x, y),$$

where the product is over all non-separable subgraphs Λ of Γ which have more than one edge (Tutte 1967).

11d *Whitney's theorem on counting subgraphs*   In Chapter 10 we obtained a formula for the coefficients of the chromatic polynomial which involved counting all the subgraphs. In this chapter we have shown that, in theory, only the non-separable subgraphs are needed. Whitney (1932b) obtained this result in a different way, by showing that there is a general expression for the number of subgraphs of any particular type in terms of the numbers of non-separable subgraphs. Specifically, let $n_t(\Gamma)$ be the number of subgraphs of Γ which have a given type t, where a 'type' is determined by the number of blocks of each isomorphism class. Then there is a polynomial function Φ_t, independent of Γ, with rational coefficients and no constant term, such that

$$n_t(\Gamma) = \Phi_t(n_\sigma(\Gamma), n_\tau(\Gamma), \dots \ ),$$

where $\sigma, \tau, \dots$ are the nonseparable types with not more edges than t. For example, if $\triangle\|$ denotes the type with one block isomorphic to K_3 and two blocks isomorphic to K_2 we have

$$n_{\triangle\|} = 6n_\triangle - \frac{7}{2} n_| n_\triangle - 2n_\theta + \frac{1}{2} n_|^2 n_\triangle,$$

where θ is the type of the theta graph $\Theta_{2,2,1}$ and the other notation is self-explanatory.

11e *An algebraic framework*   In order to unify the theory of the multiplicative expansion and Whitney's theorem described above, Biggs (1977b, 1978) introduced the following algebraic framework. Define St, the set of *star types*, to be the set of isomorphism classes of non-separable graphs, and Gr, the set of *graph types*, to be the set of functions from St to the non-negative integers, with finite support. Let $\mathbf{X}$ and $\mathbf{Y}$ respectively be the vector spaces of real-valued functions defined on St and Gr. When St is regarded as a subset of Gr in the obvious way, we have a projection $J : \mathbf{Y} \to \mathbf{X}$.

For a given graph Γ, of type g, define $\mathbf{c}_g \in \mathbf{Y}$ by the rule that $\mathbf{c}_g(t)$ is the number of subgraphs of Γ which are of type t. Then $J\mathbf{c}_g$ represents the numbers of non-separable subgraphs of Γ. Whitney's theorem asserts that there is an operator $W : \mathbf{X} \to \mathbf{Y}$, such that

$$W(J\mathbf{c}_g) = \mathbf{c}_g \quad \text{for all} \quad g \in Gr.$$

In the papers quoted, it is proved that $W = B^{-1}U$, where B is a linear operator defined by a certain infinite matrix, and $U : \mathbf{X} \to \mathbf{Y}$ is the 'monomial' mapping defined by

$$(U\mathbf{x})(t) = \prod_{\sigma \in St} \mathbf{x}(\sigma)^{t(\sigma)}.$$

11f *Expansions as linear functionals*   Denote the subspaces of $\mathbf{X}$ and $\mathbf{Y}$ consisting of vectors with finite support by $\mathbf{X}_0$ and $\mathbf{Y}_0$ respectively. The real vector spaces $\mathbf{X}_0$ and $\mathbf{Y}_0$ admit scalar products defined in the usual way:

$$\langle\!\langle \mathbf{x}_1, \mathbf{x}_2 \rangle\!\rangle = \sum_\sigma \mathbf{x}_1(\sigma)\mathbf{x}_2(\sigma); \quad \langle \mathbf{y}_1, \mathbf{y}_2 \rangle = \sum_t \mathbf{y}_1(t)\mathbf{y}_2(t).$$

For any given $\mathbf{m} \in \mathbf{Y}_0$ there is a linear functional $\mathcal{M}$ defined by $\mathcal{M}(\mathbf{y}) = \langle \mathbf{y}, \mathbf{m} \rangle$. On vectors $\mathbf{c}_g$ representing 'real graphs', $\mathcal{M}(\mathbf{c_g})$ is, by definition of the scalar product, a sum over subgraphs in which each subgraph of type t contributes $\mathbf{m}(t)$. In the author's papers quoted above it is shown that under certain conditions there is a corresponding linear functional $\mathcal{L}$ on $\mathbf{X}_0$ such that

$$\exp \mathcal{L}(J\mathbf{c}_g) = \mathcal{M}(\mathbf{c}_g) \quad \text{for all} \quad g \in Gr.$$

Explicitly, we have

$$\mathcal{L}(\mathbf{x}) = \langle\!\langle \mathbf{x}, \mathbf{l} \rangle\!\rangle, \quad \text{where} \quad \mathbf{l} = J(B^{-1})^t \mathbf{m}.$$

11g *The Hopf algebra framework*   There is clearly a substantial amount
of algebraic structure underlying Whitney's theorem and the multiplicative expansion. Schmitt (1993) carries this idea to its logical conclusion
by introducing coalgebras and Hopf algebras. He shows that the algebra
of formal power series with rational coefficients over St can be given the
structure of a Hopf algebra, and that it is isomorphic to the dual of the
free module with rational coefficients over Gr. Whitney's theorem is a
direct consequence of the isomorphism.

Another approach using Hopf algebras is discussed by Ray (1992).

12

The induced subgraph expansion

In this chapter we shall modify the multiplicative expansion of the chromatic polynomial in such a way that the induced subgraphs are the only ones occurring in the formula. This procedure has two advantages. First, there are fewer induced subgraphs than subgraphs in general; and secondly, the function which takes the place of the q function (in the notation of Proposition 11.6) turns out to be trivial for a wider class of graphs.

The formal details of the transition to induced subgraphs are quite straightforward. For any non-separable graph Λ define

$$Q(\Lambda; u) = \prod q(\Delta; u),$$

where the product is over the set of spanning subgraphs Δ of Λ, that is, those for which $V\Delta = V\Lambda$. It follows immediately that Q is a rational function of u. For example, the cycle graph C_n has just one non-separable spanning subgraph, which is C_n itself. Thus the definition of Q gives

$$Q(C_n; u) = q(C_n; u) = 1 + \frac{(-1)^n}{(u-1)^{u-1}}.$$

Proposition 12.1 *The chromatic polynomial has a multiplicative expansion*

$$C(\Gamma; u) = u^{|V\Gamma|}(1 - u^{-1})^{|E\Gamma|} \prod Q(\Lambda; u),$$

where the product is over all non-separable induced subgraphs of Γ having more than one edge.

Proof The factors which appear in Proposition 11.6 can be grouped in such a way that each group contains those subgraphs of Γ which have a given vertex-set. This grouping of factors corresponds precisely to that given in the definition of Q and the resulting expression for C, since each subgraph Δ of Γ is a subgraph of exactly one induced subgraph Λ of Γ (the one for which $V\Lambda = V\Delta$), and conversely, each subgraph of Λ is a subgraph of Γ. $\square$

The crucial fact which makes the multiplicative expansion useful in practice is that the q and Q functions are rational functions of a special kind. Specifically, it can be shown that

$$q(\Gamma; u) = 1 + \frac{\nu(u)}{\delta(u)},$$

where ν and δ are polynomials whose degrees satisfy

$$\deg \delta - \deg v \geq |V\Gamma| - 1.$$

The first satisfactory proof of this important fact was given by Tutte (1967), using the notion of 'tree mappings'. An algebraic proof was given by Biggs (1978, see also **11e** and **11f**).

Given this result, we can prove the same thing for Q.

Proposition 12.2 *Let Γ be a non-separable graph. Then $Q(\Gamma; u)$ may be written in the form*

$$Q(\Gamma; u) = 1 + \frac{\nu(u)}{\delta(u)},$$

where ν and δ are polynomials such that $\deg \delta - \deg \nu \geq |V\Gamma| - 1$.

Proof The function Q is defined to be the product of functions q, over a set of graphs with the same number of vertices. Thus the result for q implies the result for Q. $\square$

We are now in a position to overcome the circularity mentioned at the end of the previous chapter. It is possible, using Proposition 12.2, to calculate both $C(\Gamma; u)$ and $Q(\Gamma; u)$ provided only that we know the Q functions for all *proper* induced subgraphs of Γ, that is, the induced subgraphs not including Γ itself. To see this we write the formula of Proposition 12.1 as

$$C(\Gamma; u) = P(u)Q(\Gamma; u)$$

where $P(u)$ is a product of rational functions corresponding to the proper induced subgraphs, including the vertices (for each of which we have factor u) and the edges (for each of which we have a factor $1 - u^{-1}$).

It follows that $P(u)$ can be written as a polynomial of degree $n = |V\Gamma|$ plus a power series in u^{-1}:

$$P(u) = u^n + \alpha_1 u^{n-1} + \ldots + \alpha_{n-1} u + \alpha_n + \alpha_{n+1} u^{-1} + \ldots .$$

But, following Proposition 12.2, the function $Q(\Gamma; u)$ can be written

$$Q(\Gamma; u) = 1 + \beta_0 u^{-n+1} + \beta_1 u^{-n} + \ldots .$$

It follows that multiplying $P(u)$ by this expression does not alter the coefficients of $u^n, u^{n-1}, \ldots, u^2$ in $P(u)$. Thus the polynomial part of $P(u)$ is a correct expression for $C(\Gamma; u)$, except for the coefficients of u and 1. But these coefficients in $C(\Gamma; u)$ are easily found, by noting that $u(u-1)$ is a factor of $C(\Gamma; u)$. It follows that both $C(\Gamma; u)$ and $Q(\Gamma; u)$ are determined by the known function $P(u)$.

An example will elucidate this argument. Take $\Gamma = K_4$; then the only proper induced subgraphs of Γ having more than one edge are the four copies of $K_3 = C_3$. Thus

$$C(K_4; u) = u^4 (1 - u^{-1})^6 \left(1 - \frac{1}{(u-1)^2}\right)^4 Q(K_4; u)$$

$$= \frac{u^2 (u-2)^4}{(u-1)^2} Q(K_4; u).$$

Dividing $(u-1)^2$ into $u^2(u-2)^4$ gives $P(u) = u^4 - 6u^3 + 11u^2 - \ldots$, and so

$$C(K_4; u) = u^4 - 6u^3 + 11u^2 - au + b.$$

Since $u(u-1)$ is a factor of $C(K_4; u)$ it follows that $a = 6$, $b = 0$ and

$$C(K_4; u) = u^4 - 6u^3 + 11u^2 - 6u = u(u-1)(u-2)(u-3).$$

We can also find $Q(K_4; u)$ by substituting back, obtaining

$$Q(K_4; u) = 1 - \frac{2u-3}{u(u-2)^3}.$$

The technique which we have just described has the important consequence that we can calculate chromatic polynomials merely by counting induced subgraphs, without knowing any C and Q functions in advance. In particular, it implies that the chromatic polynomial is *reconstructible*, in the sense of **7f**.

To make this explicit, suppose that $\Lambda_1, \Lambda_2, \ldots, \Lambda_l$ is a list of the isomorphism types of non-separable induced subgraphs of Γ, where $K_1 = \Lambda_1$ and $K_2 = \Lambda_2$ are included, for the sake of uniformity, and $\Gamma = \Lambda_l$. Then we define a matrix $\mathbf{N} = (n_{ij})$, by putting n_{ij} equal to the number of induced subgraphs of Λ_i which are isomorphic with Λ_j. We may suppose that the list has been ordered in such a way that $\mathbf{N}$ is a triangular matrix each of whose diagonal entries is $+1$.

Proposition 12.3   *The matrix* **N** *completely determines the chromatic polynomial of* Γ.

Proof   We know the C and Q functions for all the graphs with at most three vertices. Now, suppose we know the C and Q functions for the induced subgraphs of Γ with at most t vertices; then we can find the C and Q functions for each induced subgraph with $t+1$ vertices, by using the technique previously explained. Thus, using this procedure recursively leads to the chromatic polynomial of Γ.   □

For example, the following is a complete list of the non-separable isomorphism types of induced subgraphs of the ladder graph L_3. (The graph itself occurs as Λ_6 in Figure 6.)

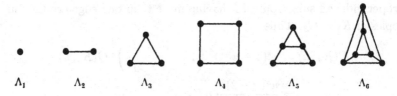

$$\Lambda_1 \qquad \Lambda_2 \qquad \Lambda_3 \qquad \Lambda_4 \qquad \Lambda_5 \qquad \Lambda_6$$

Figure 6: the induced subgraphs of L_3.

The **N** matrix for Γ is:

$$\begin{bmatrix} 1 & & & & & \\ 2 & 1 & & & & \\ 3 & 3 & 1 & & & \\ 4 & 4 & 0 & 1 & & \\ 5 & 6 & 1 & 1 & 1 & \\ 6 & 9 & 2 & 3 & 6 & 1 \end{bmatrix}.$$

To see how the method works, suppose that we have completed the calculations for subgraphs with at most four vertices. The C and Q functions for these graphs are as follows.

	Λ_2	Λ_3	Λ_4
$C:$	$u(u-1)$	$u(u-1)(u-2)$	$u(u-1)(u^2-3u+3)$
$Q:$	$(u-1)/u$	$u(u-2)/(u-1)^2$	$u(u^2-3u+3)/(u-1)^3$

The remainder of the calculation now proceeds in the following way. We have $C(\Lambda_5; u) = P_5(u) Q(\Lambda_5; u)$, where

$$P_5(u) = u^5 \left(\frac{u-1}{u} \right)^6 \left(\frac{u(u-2)}{(u-1)^2} \right) \left(\frac{u(u^2-3u+3)}{(u-1)^3} \right)$$

$$= u(u-1)(u-2)(u^2-3u+3).$$

Here (atypically) $P_5(u)$ is a polynomial divisible by $u(u-1)$, and so

$$C(\Lambda_5; u) = u(u-1)(u-2)(u^2-3u+3) \quad \text{and} \quad Q(\Lambda_5; u) = 1.$$

At the next stage, we have $C(\Lambda_6; u) = P_6(u)Q(\Lambda_6; u)$, where

$$P_6(u) = u^6 \left(\frac{u-1}{u}\right)^9 \left(\frac{u(u-2)}{(u-1)^2}\right)^2 \left(\frac{u(u^2-3u+3)}{(u-1)^3}\right)^3 (1)^6 \tag{1}$$

$$= u^6 - 9u^5 + 34u^4 - 67u^3 + 67u^2 - \ldots .$$

Here $P_6(u)$ is not a polynomial. Extending the terms in u^2 and above to a polynonial divisible by $u(u-1)$ we get $C(\Gamma; u) = u^6 - 9u^5 + 34u^4 - 67u^3 + 67u^2 - 26u$.

One noteworthy feature of the preceding calculation is that $Q(\Lambda_5; u) = 1$, although Λ_5 is a non-separable graph. This means that we could have ignored Λ_5 completely, both in setting up the matrix **N** and in the subsequent calculations. The next proposition shows that there is a large class of non-separable graphs Γ for which $Q(\Gamma; u) = 1$.

Proposition 12.4 (Baker 1971)   *If the graph Γ is quasi-separable, in the sense of Definition 9.8, then $Q(\Gamma; u) = 1$.*

Proof   We prove this result by induction on the number of vertices of Γ. The result is true for all quasi-separable graphs with at most four vertices. For this set contains only one graph (the graph shown in Fig.3, p. 67) which is not in fact separable, and the claim can be readily checked for that graph.

Suppose that the result is true for all quasi-separable graphs with at most L vertices, and let Γ be a quasi-separable graph with $L+1$ vertices. We have a quasi-separation (V_1, V_2) of Γ, where $\langle V_1 \cap V_2\rangle$ is complete and $\langle V\Gamma - (V_1\cap V_2)\rangle$ is disconnected. The expansion of Proposition 12.1 can be written in the form

$$C(\Gamma; u) = P(u)Q(\Gamma; u),$$

where $P(u)$ is a product of factors corresponding to the proper non-separable induced subgraphs of Γ. If U is any proper subset of $V\Gamma$ for which $U \not\subseteq V_1$ and $U \not\subseteq V_2$, then $\langle U\rangle$ is a quasi-separable graph, with quasi-separation $(V_1 \cap U), (V_2 \cap U)$. By the induction hypothesis, $Q(\langle U\rangle; u) = 1$.

Thus the non-trivial terms in the product $P(u)$ correspond to the subsets of V_1 and the subsets of V_2. However, a subset of $V_1 \cap V_2$ occurs just once, rather than twice. It follows that

$$P(u) = \frac{C(\langle V_1\rangle; u)\, C(\langle V_2\rangle; u)}{C(\langle V_1 \cap V_2\rangle; u)}.$$

Since Proposition 9.9 tells us that $C(\Gamma; u)$ is also equal to this expression, it follows that $Q(\Gamma; u) = 1$, and the induction step is verified.   □

We observe that the graph Λ_5, in the example preceding the proposition, is in fact quasi-separable, and so the fact that $Q(\Lambda_5; u) = 1$ is explained.

The following theorem is the culmination of the theory developed in Chapters 10–12.

Theorem 12.5   *The chromatic polynomial of a graph is determined by its proper induced subgraphs which are not quasi-separable.*

Proof This theorem follows from Propositions 12.1 and 12.4.   □

We close this chapter with a brief explanation of how the theory can be used to study the 'chromatic polynomial' of an infinite graph. Suppose Ψ is an infinite graph which can be regarded in some way as the limit of a sequence of finite graphs Ψ_n, with $|V\Psi_n| = v_n$ say. The appropriate definition of the 'chromatic polynomial' of Ψ is

$$C_\infty(\Psi; u) = \lim_{n \to \infty} \{C(\Psi_n; u)\}^{1/v_n},$$

provided the limit exists for a suitable range of values of u. In theoretical physics this is known as taking the *thermodynamic limit*, and some existence results have been proved for 'interaction models', as defined in **10f**. Grimmett (1978) obtained strong results for the rank polynomial, but for our present purposes, blind faith and ignorance will suffice.

If Ψ_n has reasonable regularity properties, then the number of induced subgraphs of a given type in Ψ_n is αv_n, where α is a constant representing the *density*, that is the number of induced subgraphs of that type per vertex. For example, if Ψ_n is regular of degree k, the number of edges is $(k/2)v_n$, and so the density of edges is $k/2$. If we now take the $(1/v_n)$th root of the multiplicative Q-formula for $C(\Psi_n; u)$ we get a term u (corresponding to the vertices), a term $(1 - u^{-1})^{k/2}$ (corresponding to the edges), and in general a term $Q(\Lambda; u)^\alpha$ for each induced subgraph Λ of density α. This leads to a definition of the 'chromatic polynomial' which does not depend on the approximating sequence Ψ_n. Unfortunately, nothing is known about the convergence of the infinite product, although it is clear that the smallest induced subgraphs, which are the easiest to count, contribute the largest terms.

A good illustration is provided by the infinite plane square lattice graph. Here the only induced subgraphs which are not quasi-separable and have not more than eight vertices are: the vertices, edges, C_4's and C_8's, with densities $1, 2, 1$ and 1 respectively. It follows that an approximation to C_∞ in this case is

$$u(1 - u^{-1})^2 \left(1 + \frac{1}{(u-1)^3}\right) \left(1 + \frac{1}{(u-1)^7}\right).$$

The correct value when $u = 3$ is known to be $(4/3)^{3/2} = 1.540\ldots$ (Lieb 1967), whereas the approximation gives $1.512\ldots$. For larger values of u it seems likely that the approximation is better, but no general results are known. (See also **12f**.)

Additional Results

12a *The Q function for complete graphs*   We have

$$Q(K_n; u) = \prod_{0 \le i \le n-1} (u - i)^{\eta(i)},$$

where $\eta(i) = (-1)^{n-1-i}\binom{n-1}{i}$.

12b *The Q functions for all graphs with less than six vertices*   The only graphs with less than five vertices which are not quasi-separable are K_2, K_3, K_4, and C_4, and we have already found Q for all these. Writing $Q(\Gamma; u) = 1 + r(\Gamma; u)$, the r functions are as follows:

$$
\begin{aligned}
r(K_2; u) &= 1/u \\
r(K_3; u) &= -1/(u-1)^2 \\
r(K_4; u) &= -(2u-3)/u(u-2)^2 \\
r(C_4; u) &= 1/(u-1)^3.
\end{aligned}
$$

The relevant graphs with five vertices are: K_5, W_5, W_5^- (the wheel with one 'spoke' removed), $K_{2,3}$ and C_5. The r functions are:

$$
\begin{aligned}
r(K_5; u) &= -(6u^4 - 48u^3 + 140u^2 - 176u + 81)/(u-1)^4(u-3)^4 \\
r(W_5; u) &= (3u^2 - 9u + 7)/u(u-2)^3(u^2 - 3u + 3) \\
r(W_5^-; u) &= (2u^2 - 6u + 5)/u(u-2)(u^2 - 3u + 3)^2 \\
r(K_{2,3}; u) &= (u^3 - 6u^2 + 11u - 7)/u(u^2 - 3u + 3)^3 \\
r(C_5; u) &= -1/(u-1)^4.
\end{aligned}
$$

12c *Petersen's graph*   The only non-quasi-separable induced subgraphs of Petersen's graph O_3 have $2, 5, 6, 7, 8, 9, 10$ vertices, and there is one isomorphism class in each case. The $\mathbf{N}$ matrix is

$$\begin{bmatrix} 1 & & & & & & \\ 5 & 1 & & & & & \\ 6 & 0 & 1 & & & & \\ 8 & 2 & 1 & 1 & & & \\ 10 & 4 & 2 & 4 & 1 & & \\ 12 & 6 & 4 & 9 & 3 & 1 & \\ 15 & 12 & 10 & 30 & 15 & 10 & 1 \end{bmatrix}.$$

Using the method described on pp. 92–93 this gives the chromatic polynomial of O_3:

$$u(u-1)(u-2)(u^7 - 12u^6 + 67u^5 - 230u^4 + 529u^3 - 814u^2 + 775u - 352).$$

12d *The first non-trivial coefficient in q and Q*   If Γ is non-separable and has n vertices and m edges, then the coefficient of $u^{-(n-1)}$ in the expression for $q(\Gamma; u)$ in descending powers of u is equal to $(-1)^m$. The corresponding coefficient in $Q(\Gamma; u)$ is therefore $\sum (-1)^{|E\Lambda|}$, where the summation is over all non-separable spanning subgraphs Λ of Γ (Tutte 1967).

12e *Chromatic powers*   Let $\sigma_m(\Gamma)$ denote the sum of the mth powers of the zeros of $C(\Gamma; u)$. Suppose that

$$\log Q(\Lambda; u) = -\sum_{j=1}^{\infty} \frac{c_j(\Lambda)}{ju^j},$$

where the expansion is valid for $|u|$ sufficiently large. If $n(\Gamma; \Lambda)$ denotes the number of induced subgraphs of Γ which are isomorphic with Λ, we have

$$\sigma_m(\Gamma) = \Sigma n(\Gamma; \Lambda) c_m(\Lambda),$$

where the sum is taken over isomorphism classes of non-quasi-separable graphs (Tutte 1967).

12f *Approximations for the infinite square lattice*   There have been many attempts to determine the 'chromatic polynomial' $C_\infty(u)$ of the infinite square lattice. Biggs and Meredith (1976) obtained the estimate

$$\frac{1}{2}(u - 3 + \sqrt{u^2 - 2u + 5}).$$

Using the 'transfer matrix' method, Biggs (1977a) obtained the bounds

$$\frac{u^2 - 3u + 3}{u - 1} \leq C_\infty(u) \leq \frac{1}{2}(u - 2 + \sqrt{u^2 - 4u + 8}).$$

Kim and Enting (1979) obtained a series approximation in terms of $x = 1/(u-1)$: apart from a simple factor it is

$$1 + x^3 + x^7 + 3x^8 + 4x^9 + 3x^{10} + 3x^{11} + 11x^{12} + 24x^{13} + 8x^{14}$$

$$- 91x^{15} - 261x^{16} - 290x^{17} + 254x^{18} + \dots .$$

13

The Tutte polynomial

There is a remarkable relationship between the rank polynomial and the spanning trees of a graph. In this chapter we shall develop this theory by giving an explicit definition of what is known as the *Tutte polynomial* $T(\Gamma; x, y)$ of a graph Γ in terms of its spanning trees, and then proving an identity between the Tutte polynomial and the rank polynomial.

An alternative approach to the Tutte polynomial is to define it recursively by the deletion–contraction property,

$$T(\Gamma; x, y) = T(\Gamma^{(e)}; x, y) + T(\Gamma_{(e)}; x, y),$$

where e is neither a loop nor an isthmus. This rule, together with a 'boundary condition' (see **13c**), does in fact define T completely. However, it is not immediately obvious that the method leads to a result which is independent of the order in which edges are deleted and contracted, and it provides no insight into the remarkable properties of T. For these reasons we shall follow the constructive route given below.

The definition of the rank polynomial depends upon the assignment of the ordered pair (*rank*, *co-rank*) of non-negative integers to each subgraph; we shall call such an assignment a *bigrading* of the set of subgraphs. If Γ is connected, the set of subgraphs whose bigrading is $(r(\Gamma), 0)$ is just the set of spanning trees of Γ. We shall introduce a new bigrading of subgraphs which has the property that, if it is given only for the spanning trees of Γ, then the entire rank polynomial of Γ is determined. Our procedure is based initially upon an ordering of the edge-set $E\Gamma$, although a consequence of our main result is the fact that this arbitrary ordering is essentially irrelevant. Another consequence of the main result is an expansion of the chromatic polynomial in terms of spanning trees; this will be the subject of Chapter 14.

We now fix some hypotheses and conventions which will remain in force throughout this chapter. The graph Γ is a connected general graph, and $E\Gamma$ has a fixed total ordering denoted by $\leq$. If $X \subseteq E\Gamma$ we shall use the symbol X (rather than $\langle X \rangle$) to denote the corresponding edge-subgraph of Γ, and the singleton sets $\{x\} \subseteq E\Gamma$ will be denoted by x instead of $\{x\}$. The rank of Γ will be denoted by r_0; thus $r_0 = r(\Gamma) = |V\Gamma| - 1$.

If $X \subseteq E\Gamma$ and $x \in E\Gamma \setminus X$, then the rank of $X \cup x$ is either $r(X)$ or $r(X) + 1$, and in the latter case we say that x is *independent* of X. Now if $r(X) \neq r_0$, there will certainly be some edges of Γ which are independent of X, and we shall denote the first of these (in the ordering $\leq$) by $\lambda(X)$. We note that, since

$$r(Y) + s(Y) = |Y| \quad \text{for all} \quad Y \subseteq E\Gamma,$$

we have the equations

$$r(X \cup \lambda(X)) = r(X) + 1, \quad s(X \cup \lambda(X)) = s(X).$$

Similarly, if $s(X) \neq 0$ then there are some edges x for which $s(X \setminus x) = s(X) - 1$, and we denote the first of these by $\mu(X)$. We have

$$r(X \setminus \mu(X)) = r(X), \quad s(X \setminus \mu(X)) = s(X) - 1.$$

Definition 13.1   The λ *operator* on subsets of $E\Gamma$ assigns to each set $X \subseteq E\Gamma$ the set X^λ derived from X by successively adjoining the edges $\lambda(X), \lambda(X \cup \lambda(X)), \ldots$, until no further increase in the rank is possible. The μ *operator* takes X to the set X^μ, which is derived from X by successively removing the edges $\mu(X), \mu(X \setminus \mu(X)), \ldots$, until no further decrease in the co-rank is possible.

We notice the following properties of the λ and μ operators:

$$X \subseteq X^\lambda, \quad r(X^\lambda) = r_0, \quad s(X^\lambda) = s(X).$$

$$X^\mu \subseteq X, \quad r(X^\mu) = r(X), \quad s(X^\mu) = 0.$$

We shall exploit the obvious similarity between the two operators by giving proofs only for one of them. The first lemma says that the edges which must be added to a subgraph A to form A^λ can be added in any order. (In what follows the notation $x < y$ will mean $x \leq y$ and $x \neq y$.)

Lemma 13.2   If $A \subseteq B \subseteq A^\lambda$, then $B^\lambda = A^\lambda$.

Proof   If $A = A^\lambda$, the result is trivial. Suppose

$$A^\lambda \setminus A = X = \{x_1, x_2, \ldots, x_t\},$$

where $x_1 < x_2 < \ldots < x_t$, and let $B = A \cup Y$, where $Y \subseteq X$. If $Y = X$, then $B = A^\lambda$ and $B^\lambda = A^{\lambda\lambda} = A^\lambda$. If $Y \neq X$, let x_a be the first edge in $X \setminus Y$. Then, if an edge x is independent of B, it follows that x is independent of $A \cup \{x_1, \ldots, x_{a-1}\}$ (which is contained in B), and so $x_a \leq x$, since x_a is the first edge independent of

$$A \cup \{x_1, \ldots, x_{a-1}\}.$$

But x_a itself is certainly independent of B, since when we add the edges in X to A, the rank must increase by exactly one at each step. Thus $x_a = \lambda(B)$, and by successively repeating the argument with $B' = B \cup \lambda(B), B'' = B' \cup \lambda(B'), \ldots$, we have the result. $\qquad\square$

Lemma 13.3   *If $A \subseteq B$ and $r(B) \neq r_0$, then $\lambda(B) \in A^\lambda$.*

Proof   Since $r(B) \neq r_0$, there is a first edge $\lambda(B)$ independent of B, and consequently independent of A. Suppose $\lambda(B)$ is not in A^λ. Then each edge x in $A^\lambda \setminus A$ must satisfy $x < \lambda(B)$, and so x is not independent of B; also, since $A \subseteq B$, no edge in A is independent of B. Thus all edges in A^λ are not independent of B, and $r(B) = r(A^\lambda) = r_0$. This is a contradiction, so our hypothesis was false and $\lambda(B)$ is in A^λ. $\qquad\square$

We note the analogous properties of the μ operator:

$$A^\mu \subseteq B \subseteq A \implies B^\mu = A^\mu; \quad B \subseteq A \quad \text{and} \quad s(B) \neq 0 \implies \mu(B) \notin A^\mu.$$

The next definition introduces a new bigrading of the subsets of $E\Gamma$.

Definition 13.4   Let X be a subset of $E\Gamma$. An edge e in $E\Gamma \setminus X$ is said to be *externally active* with respect to X if $\mu(X \cup e) = e$. An edge f in X is said to be *internally active* with respect to X if $\lambda(X \setminus f) = f$. The number of edges which are externally (internally) active with respect to X is called the *external (internal) activity* of X.

We shall denote the sets of edges which are externally and internally active with respect to X by X^ϵ and X^ι respectively, and use the notation

$$X^+ = X \cup X^\epsilon, \quad X^- = X \setminus X^\iota.$$

These concepts are motivated by their interpretation in the case of a spanning tree, because in that case they are related to the systems of basic cycles and cuts which were discussed in Chapter 5.

Proposition 13.5   *For any spanning tree T of Γ we have:*
(1) the edge e is externally active with respect to T if and only if e is the first edge (in the ordering $\leq$) of $\mathrm{cyc}(T, e)$;
(2) the edge f is internally active with respect to T if and only if f is the first edge (in the ordering $\leq$) of $\mathrm{cut}(T, f)$.

Proof By definition, e is externally active if and only if e is the first edge whose removal decreases the co-rank of $T \cup e$. But $T \cup e$ contains just one cycle, which is $\text{cyc}(T, e)$, and any edge whose removal decreases the co-rank must belong to this cycle.

The second part is proved by a parallel argument. $\quad\square$

Definition 13.6 The *Tutte polynomial* of a connected graph Γ, with respect to an ordering $\leq$ of $E\Gamma$, is defined as follows. Suppose t_{ij} is the number of spanning trees of Γ whose internal activity is i and whose external activity is j. Then the Tutte polynomial is

$$T(\Gamma, \leq; x, y) = \sum t_{ij} x^i y^j.$$

Remarkably, it will turn out that T is independent of the chosen ordering.

In order to obtain the main result we shall investigate the relationship between the concepts just defined and the following diagram of operators.

$$
\begin{array}{ccc}
\mathcal{A} & \xrightarrow{\lambda} & \mathcal{B} \\
\downarrow{\mu} & & \downarrow{\mu} \\
\mathcal{C} & \xrightarrow{\lambda} & \mathcal{D}
\end{array}
$$

Here $\mathcal{A}$ denotes all subsets of $E\Gamma$, $\mathcal{B}$ denotes subsets Z with $r(Z) = r_0$, $\mathcal{C}$ denotes subsets W with $s(W) = 0$, and $\mathcal{D}$ denotes subsets T with $r(T) = r_0$ and $s(T) = 0$ (that is, spanning trees). It is worth remarking that the diagram is commutative, although we shall not need this result (see **13g**).

Proposition 13.7 *Let X be any subset in the image of the λ operator, so that $r(X) = r_0$ and $X^\lambda = X$. Then*

$$X = Y^\lambda \iff X^- \subseteq Y \subseteq X.$$

Proof Suppose $X = Y^\lambda$. Then $Y \subseteq Y^\lambda = X$, so $Y \subseteq X$. If f is an edge of X^-, then certainly f is in $X = Y^\lambda$. If f were in $Y^\lambda \setminus Y$, then by Lemma 13.2, $\lambda(Y^\lambda \setminus f) = f$; but this means that f is internally active with respect to $X = Y^\lambda$, contradicting $f \in X^-$. Thus f is in Y, and $X^- \subseteq Y$.

Suppose $X^- \subseteq Y \subseteq X$. If $X = Y$, then we have $X = X^\lambda = Y^\lambda$. Now if $f \in X \setminus Y$, then f is internally active with respect to X and so $\lambda(X \setminus f) = f$. From $Y \subseteq X \setminus f$ we have (by Lemma 13.3) $\lambda(X \setminus f) \in Y^\lambda$, that is, $f \in Y^\lambda$. Since this is true for all f in $X \setminus Y$, it follows that $X \setminus Y \subseteq Y^\lambda$, and consequently $X \subseteq Y^\lambda$. Finally, from Definition 13.1 and $Y \subseteq X \subseteq Y^\lambda$ we deduce that $X^\lambda = Y^\lambda$, that is $X = Y^\lambda$. $\quad\square$

We note the analogous result: if X is in the image of the μ operator, then

$$X = Y^\mu \iff X \subseteq Y \subseteq X^+.$$

Proposition 13.8   *Let T be a spanning tree (that is, $T \in \mathcal{D}$), and suppose $W \in \mathcal{C}$ is such that $W^\lambda = T$. Then $W^\epsilon = T^\epsilon$.*

Proof   Suppose that the edge e is externally active with respect to T. We shall show that the whole of $\text{cyc}(T, e)$ belongs to W, whence it follows that e is externally active with respect to W. If there is an edge $f \neq e$ in $\text{cyc}(T, e)$ which is not in W, then, since (by Proposition 13.7) we have $T^- \subseteq W \subseteq T$, f must be internally active with respect to T. Now $f \in \text{cyc}(T, e)$ implies that $e \in \text{cut}(T, f)$, and the internally active property of f means that $f < e$. This contradicts the externally active property of e. Hence $\text{cyc}(T, e) \subseteq W$, and e is externally active with respect to W.

Conversely, if e is externally active with respect to W it follows immediately that e is externally active with respect to T.   □

We now set up the main theorem, using the portion $\mathcal{A} \xrightarrow{\mu} \mathcal{C} \xrightarrow{\lambda} \mathcal{D}$ of the operator diagram. Define

$$\tilde{\rho}_{ij} = |\{X \in \mathcal{A} \mid r(X) = r_0 - i,\ s(X) = j\}|,$$
$$\pi_{ij} = |\{W \in \mathcal{C} \mid r(W) = r_0 - i,\ |W^\epsilon| = j\}|,$$
$$t_{ij} = |\{T \in \mathcal{D} \mid |T^\iota| = i,\ |T^\epsilon| = j\}|.$$

Of course, the last line merely repeats Definition 13.6. We have three corresponding two-variable polynomials.

$$\tilde{R}(\Gamma; x, y) = \sum \tilde{\rho}_{ij} x^i y^j, \quad P(\Gamma, \leq; x, y) = \sum \pi_{ij} x^i y^j,$$
$$T(\Gamma, \leq; x, y) = \sum t_{ij} x^i y^j,$$

where the *modified rank polynomial* $\tilde{R}$ is related to the usual one (Definition 10.1) by $\tilde{R}(\Gamma; x, y) = x^{r_0} R(\Gamma; x^{-1}, y)$.

Theorem 13.9   *Let Γ be a connected graph with n vertices and let $\leq$ be any ordering of $E\Gamma$. Then the Tutte polynomial is related to the rank polynomial as follows:*

$$T(\Gamma, \leq; x + 1, y + 1) = \tilde{R}(\Gamma; x, y) = x^{n-1} R(\Gamma; x^{-1}, y).$$

Proof   We shall use the intermediate polynomial P defined above, and prove the equalities

$$T(\Gamma, \leq; x + 1, y + 1) = P(\Gamma, \leq; x, y + 1) = \tilde{R}(\Gamma; x, y),$$

which are equivalent to the following relationships among the coefficients:

$$\pi_{ij} = \sum_k \binom{k}{i} t_{kj}, \quad \tilde{\rho}_{ij} = \sum_l \binom{l}{j} \pi_{il}.$$

For the first identity, consider $\lambda : \mathcal{C} \to \mathcal{D}$. By Proposition 13.7, if T is in $\mathcal{D}$, then

$$T = W^\lambda \quad \text{if and only if} \quad T^- \subseteq W \subseteq T.$$

Also, by Proposition 13.8, the external activities of T and W are the same. Consequently, for each one of the t_{kj} spanning trees T with $|X^\iota| = k$ and $|X^\epsilon| = j$ there are $\binom{k}{i}$ subgraphs W in $\mathcal{C}$ with $r(W) = r_0 - i$ and $|W^\epsilon| = j$. These subgraphs are obtained by removing from T any set of i edges contained in the k internally active edges of T. This proves the first identity.

For the second identity, we consider $\mu : \mathcal{A} \to \mathcal{C}$. By the analogue of Proposition 13.7 for μ, if X is in $\mathcal{C}$, then

$$X = Y^\mu \quad \text{if and only if} \quad X \subseteq Y \subseteq X^+.$$

Consequently, for each one of the π_{il} subgraphs X in $\mathcal{C}$ with $r(X) = r_0 - i$ and $|X^\epsilon| = l$, there are $\binom{l}{j}$ subgraphs Y with $r(Y) = r_0 - i$ and $s(Y) = j$. These subgraphs are obtained by adding to X any set of j edges contained in the l externally active edges of X. This proves the second identity. $\qquad\square$

Corollary 13.10   *The Tutte polynomial of a connected graph Γ is independent of the ordering used in its definition.*

Proof   This statement follows from Theorem 13.9 and the fact that the rank polynomial is independent of the ordering. $\qquad\square$

The original proof of Theorem 13.9 by Tutte (1954) was inductive; the proof given above is a simplification of the first constructive proof by Crapo (1969). In the light of the Corollary we can write $T(\Gamma; x, y)$ for the Tutte polynomial of Γ. It should be noted that, although each coefficient t_{ij} is independent of the ordering, the corresponding set of spanning trees (having internal activity i and external activity j) does depend on the ordering.

Additional Results

13a *Tutte polynomials of cycles*   By listing the spanning trees of C_n and calculating their internal and external activities we obtain

$$T(C_n; x, y) = y + x + x^2 + \ldots + x^{n-1}.$$

13b *The Tutte matrix of Petersen's graph* (Biggs 1973b)   The matrix (t_{ij}) of coefficients of the Tutte polynomial for Petersen's graph is

$$\begin{bmatrix} 0 & 36 & 84 & 75 & 35 & 9 & 1 \\ 36 & 168 & 171 & 65 & 10 \\ 120 & 240 & 105 & 15 \\ 180 & 170 & 30 \\ 170 & 70 \\ 114 & 12 \\ 56 \\ 21 \\ 6 \\ 1 \end{bmatrix}.$$

13c *The deletion–contraction property*   The following two properties completely define the Tutte polynomial for connected graphs.

(1) If e is an edge of the connected graph Γ which is neither a loop nor an isthmus, then $T(\Gamma; x, y) = T(\Gamma^{(e)}; x, y) + T(\Gamma_{(e)}; x, y)$.

(2) If Λ_{ij} is formed from a tree with i edges by adding j loops, $T(\Lambda_{ij}; x, y) = x^i y^j$.

13d *Recursive families* (Biggs, Damerell and Sands 1972)   Using the deletion–contraction property we can obtain a second-order recurrence for the Tutte polynomials of the cycle graphs:

$$T(C_{n+2}; x, y) - (x + 1)T(C_{n+1}; x, y) + xT(C_n; x, y) = 0.$$

Generally, a family $\{\Gamma_l\}$ of graphs is said to be a *recursive family* if there is a linear recurrence of the form

$$T(\Gamma_{l+\rho}; x, y) + a_1 T(\Gamma_{l+\rho-1}; x, y) + \ldots + a_\rho T(\Gamma_l; x, y) = 0,$$

where the coefficients $a_1, \ldots, a_\rho$ are polynomial functions of (x, y), and are independent of l. Thus the cycle graphs form a recursive family with $\rho = 2$. The families $\{L_h\}$, $\{M_h\}$ of ladders and Möbius ladders are recursive families with $\rho = 6$; they have the same recurrence, whose auxiliary equation is

$$(t - 1)(t - x)(t^2 - (x + y + 2)t + xy)(t^2 - (x^2 + x + y + 1)t + x^2 y) = 0.$$

From this we can deduce the tree-numbers and the chromatic polynomials for these graphs. (See also **9c**.)

13e *Tutte polynomials of complete graphs*   Let $\tau(x, y, \alpha)$ and $\rho(y, \alpha)$ be the exponential generating functions for the polynomials $T(K_n; x, y)$ and $y^{\binom{n}{2}}(y - 1)^{-n}$ respectively. Then

$$\tau(x, y, \alpha) = \frac{\rho(y, \alpha)^{(x-1)(y-1)} - 1}{x - 1}.$$

13f *Inversions of trees*   A *labelled tree* on n vertices is a spanning tree A of K_n with the vertex-set $\{1, 2, \ldots, n\}$. Let $\text{inv}(A)$ denote the number of edges $\{i, j\}$ of A for which $i < j$ and j is on the path in A from 1 to i. Then we have

$$T(K_n; 1, y) = \sum_A y^{\text{inv}(A)},$$

where the sum is over all labelled trees on n vertices.

13g *The commutative diagram*   If $X \subseteq E\Gamma$, define

$$T = X^\mu \cup (X^\lambda \setminus X) = X^\lambda \setminus (X \setminus X^\mu).$$

Then $X^{\lambda\mu} = T = X^{\mu\lambda}$ (Crapo 1969).

13h *Counting forests*   If we write $T(\Gamma; 1, 1 + t) = \Sigma\phi_i t^i$, then ϕ_i is the number of forests in Γ which have $|V\Gamma| - i - 1$ edges. It follows that $T(\Gamma; 1, 2)$ is the total number of forests in Γ, and $T(\Gamma; 1, 1)$ is the tree-number of Γ.

13i *Planar graphs*   If Γ and Γ^* are dual planar graphs, then there is a bijective correspondence between their spanning trees which switches internal and external activity. It follows that $t_{ij} = t_{ji}^*$ and

$$T(\Gamma; x, y) = T(\Gamma^*; y, x).$$

13j *The medial graph*   Let Γ be a connected graph which is embedded in the plane. For each $e \in E(\Gamma)$ choose an interior point $m(e)$ on e. The *medial graph* $M(\Gamma)$ associated with the given embedding of Γ has vertex-set $\{m(e) \mid e \in E(\Gamma)\}$ and edge-set defined as follows. For each face of the embedded graph Γ there is a cycle with edges $e_1, e_2, \ldots, e_k$ bounding that face; we create a corresponding sequence $\mu_1, \mu_2, \ldots, \mu_k$ of edges of $M(\Gamma)$ which (i) forms a cycle in $M(\Gamma)$ with vertices $m(e_1), m(e_2), \ldots, m(e_k)$ and (ii) is topologically identical with the original cycle. $M(\Gamma)$ is a 4-regular graph and as such it has at least one Eulerian partition; that is, a partition of its edge-set into cycles without repeated edges. Let f_k denote the number of Eulerian partitions of $M(G)$ into k cycles such that, at any any vertex of $M(G)$, the two cycles passing through that vertex do not cross, in the obvious topological sense. Las Vergnas (1978) proved that

$$T(\Gamma; x, x) = \sum_{k \geq 0} f_{k+1}(x - 1)^k.$$

See also Jaeger (1988) and Las Vergnas (1988).

13k *Tutte polynomials for knots and links* (Thistlethwaite 1987)   A knot or link L is usually represented by a diagram in the plane; the diagram is said to be *alternating* if the 'crossings' are alternately over and under as we traverse each component. Associated with an alternating diagram is a graph D_L such that the *Jones polynomial* of L is given by

$$V_L(t) = (-t)^{-K} T(D_L; -t, -t^{-1}),$$

where K is a number depending on L.

This relationship leads to a simple proof of a conjecture made by Tait in the 19th century: the number of crossings in any alternating diagram of a given link is invariant, provided there are no 'nugatory' crossings.

13l *Intractability of calculating the Tutte polynomial*   A counting problem is said to be *#P-hard* if it has a certain technical property which, it is believed, is equivalent to computational intractability. Jaeger, Vertigan and Welsh (1990) showed that computing $T(\Gamma; x, y)$ is *#P-hard*, except for a few points and curves in the complex (x, y)-plane. In particular, computing the Jones polynomial (**13k**) of an alternating link is *#P-hard*.

14

Chromatic polynomials and spanning trees

In this chapter we shall study the relationship between the Tutte polynomial and the chromatic polynomial of a connected graph. The main result is as follows.

Theorem 14.1   *Let Γ be a connected graph with n vertices. Then*

$$C(\Gamma; u) = (-1)^{n-1} u \sum_{i=1}^{n-1} t_{i0}(1-u)^i,$$

where t_{i0} is the number of spanning trees of Γ which have internal activity i and external activity zero (with respect to any fixed ordering of $E\Gamma$).

Proof   We have only to invoke some identities derived in earlier chapters. The chromatic polynomial is related to the rank polynomial as in Corollary 10.5, and the rank polynomial is related to the Tutte polynomial as in Theorem 13.9. Thus we have

$$C(\Gamma; u) = u^n R(\Gamma; -u^{-1}, -1)$$
$$= u^n (-u^{-1})^{n-1} \widetilde{R}(\Gamma; -u, -1)$$
$$= (-1)^{n-1} u T(\Gamma; 1-u, 0).$$

The result follows from the definition of the Tutte polynomial. $\square$

This theorem indicates a purely algebraic way of calculating chromatic polynomials. If we are given the incidence matrix of a graph Γ, then the basic cycles and cuts associated with each spanning tree T of Γ can be found by matrix operations, as explained in Chapter 5. From this information we can compute the internal and external activities of

T, using the results of Proposition 13.5. The method is impracticable for hand calculation, but it is well-adapted to automatic computation in view of the availability of sophisticated programs for carrying out matrix algebra. Furthermore, it is demonstrably better than the deletion–contraction method (see **14h**).

Theorem 14.1 also has theoretical implications for the study of chromatic polynomials, and the remainder of this chapter is devoted to some of these consequences. First, we observe that if the chromatic polynomial is expressed in the 'reduced form'

$$C(\Gamma; u) = \pm w(w-1) \sum_{i=0}^{n-2} a_i w^i \quad \text{where} \quad w = 1 - u,$$

then the coefficients a_i are all non-negative. In fact a_i is the number $t_{i+1,0}$. It is convenient to use the reduced form to record chromatic polynomials, because the coefficients have fixed sign and are relatively small.

Proposition 14.2 *Let Γ be a connected graph, and let (t_{ij}) denote the matrix of coefficients of its Tutte polynomial. Then*

$$t_{10} = t_{01}.$$

Proof Suppose that the ordering of $E\Gamma = \{e_1, e_2, \ldots, e_m\}$ is the natural order of the subscripts. If T is a spanning tree with internal activity 1 and external activity 0, then e_1 must be an edge of T, otherwise it would be externally active. Further, e_2 is not an edge of T, otherwise both e_1 and e_2 would be internally active. Also e_1 is in $\mathrm{cyc}(T, e_2)$, otherwise e_2 would be externally active. Consequently $T_* = (T \setminus e_1) \cup e_2$ is a spanning tree, with internal activity 0 and external activity 1.

Reversing the argument shows that $T \mapsto T_*$ is a bijection, and hence t_{10} (the number of spanning trees T with $|T^\iota| = 1$ and $|T^\epsilon| = 0$) is equal to t_{01} (the number of spanning trees T_* with $|T_*^\iota| = 0$ and $|T_*^\epsilon| = 1$). $\qquad\square$

The number t_{10} has appeared in the work of several authors, for example Crapo (1967) and Essam (1971). We note that it is the coefficient a_0 in the reduced form of the chromatic polynomial. It is sufficiently important to warrant a name.

Definition 14.3 The *chromatic invariant* $\theta(\Gamma)$ of a connected graph Γ is the number of spanning trees of Γ which have internal activity 1 and external activity 0.

Theorem 14.1 provides another interpretation of $\theta(\Gamma)$ in terms of the chromatic polynomial of Γ. Let C' denote the derivative of C; then a simple calculation shows that

$$C'(\Gamma; 1) = (-1)^n t_{10} = (-1)^n \theta(\Gamma).$$

When Γ is non-separable it has at least one spanning tree with internal activity 1 and external activity 0 (**14b**). Thus, for a non-separable graph with an even number of vertices C is increasing at its zero $u = 1$, whereas if the graph has an odd number of vertices it is decreasing.

The link with the chromatic polynomial can also be used to justify the use of the name 'invariant' for $\theta(\Gamma)$. Recall that two graphs are said to be *homeomorphic* if they can both be obtained from the same graph by inserting extra vertices of degree two in its edges.

Proposition 14.4 *If Γ_1 and Γ_2 are homeomorphic connected graphs with at least two edges, then*

$$\theta(\Gamma_1) = \theta(\Gamma_2).$$

Proof Let Γ be a graph which has at least three edges, and a vertex of degree two. Let e and f be the edges incident with this vertex. The deletion of either e or f, say e, results in a graph $\Gamma^{(e)}$ in which the edge f is attached at a cut-vertex to a graph Γ_0 with at least one edge. Hence $C(\Gamma^{(e)}; u)$ is of the form $(u-1)C(\Gamma_0; u)$, where $C(\Gamma_0; 1) = 0$. The contraction of e in Γ results in a graph homeomorphic with Γ. We have

$$C(\Gamma; u) = C(\Gamma^{(e)}; u) - C(\Gamma_{(e)}; u)$$
$$= (u - 1)C(\Gamma_0; u) - C(\Gamma_{(e)}; u),$$

and, on differentiating and putting $u = 1$, we find

$$C'(\Gamma; 1) = -C'(\Gamma_{(e)}; 1).$$

Since Γ has one more vertex than $\Gamma_{(e)}$, it follows that

$$\theta(\Gamma) = \theta(\Gamma_{(e)}).$$

Now, if two graphs are homeomorphic, then they are related to some graph by a sequence of operations like that by which $\Gamma_{(e)}$ was obtained from Γ; hence we have the result. $\qquad\square$

It is worth remarking that both the proof and the result fail in the case where one of the graphs is K_2; we have $\theta(K_2) = 1$, whereas any path graph P_n $(n \geq 3)$ is homeomorphic with K_2 but $\theta(P_n) = 0$.

We end this chapter with an application of Theorem 14.1 to the *unimodal conjecture* of Read (1968). This is the conjecture that if

$$C(\Gamma; u) = u^n - c_1 u^{n-1} + \ldots + (-1)^{n-1} c_{n-1} u,$$

then, for some number M in the range $1 \leq M \leq n - 1$, we have

$$c_1 \leq c_2 \leq \ldots \leq c_M \geq c_{M+1} \geq \ldots \geq c_{n-1}.$$

There is strong numerical evidence to support this conjecture, but a proof seems surprisingly elusive. The following partial result was obtained by Heron (1972).

Proposition 14.5   *Using the above notation for the chromatic polynomial of a connected graph Γ with n vertices, we have*

$$c_{i-1} \leq c_i \quad \text{for all} \quad i \leq \frac{1}{2}(n - 1).$$

Proof   The result of Theorem 14.1 leads to the following expression for the coefficients of the chromatic polynomial:

$$c_i = \sum_{l=0}^{i} t_{n-1-l,0} \binom{n-1-l}{n-1-i} = \sum_{l=0}^{i} t_{n-1-l,0} \binom{n-1-l}{i-l}.$$

Now if $i \leq \frac{1}{2}(n - 1)$, then $i - l \leq \frac{1}{2}(n - 1 - l)$ for all $l \geq 0$. Hence, by the unimodal property of the binomial coefficients, we have

$$\binom{n-1-l}{i-l} \geq \binom{n-1-l}{i-1-l} \quad \text{for} \quad i \leq \frac{1}{2}(n-1), \quad l \geq 0.$$

Thus, since each number $t_{n-1-l,0}$ is a non-negative integer, it follows that $c_i \geq c_{i-1}$ for $i \leq \frac{1}{2}(n - 1)$, as required.

Additional Results

14a *A product formula for θ*   If Γ has a quasi-separation (V_1, V_2) with $|V_1 \cap V_2| = t$ then

$$\theta(\Gamma) = (t - 1)! \, \theta(\langle V_1 \rangle) \theta(\langle V_2 \rangle).$$

This formula is particularly useful when $t = 2$.

14b *Graphs with a given value of θ*   A connected graph Γ is separable if and only if $\theta(\Gamma) = 0$. It is a *series–parallel* graph if and only if $\theta(\Gamma) \leq 1$ (Brylawski 1971). One graph with $\theta = 2$ is K_4, and it follows from Brylawski's result on series–parallel graphs that if Γ contains a subgraph homeomorphic to K_4 then $\theta(\Gamma) \geq 2$. In order to show that all values of θ can occur we need only remark that for the wheel W_n we have $\theta(W_n) = n - 2$. Using the product formula **14a** we can construct infinitely many graphs with any given value of θ, by gluing any edge of any series-parallel graph to any edge of the appropriate wheel.

14c *The chromatic invariants of dual graphs*   Let Γ and Γ^* be dual planar connected graphs. Then

$$\theta(\Gamma) = \theta(\Gamma^*).$$

For instance,

$$\theta(Q_3) = \theta(K_{2,2,2}) = 11; \quad \theta(Icosahedron) = \theta(Dodecahedron) = 4412.$$

14d *Some explicit formulae*   For the complete graphs K_n, the ladders L_h, and the Möbius ladders M_h, we have

$$\theta(K_n) = (n-2)! \quad (n \geq 2),$$
$$\theta(L_h) = 2^h - h - 1 \quad (h \geq 3),$$
$$\theta(M_h) = 2^h - h \quad (h \geq 2).$$

14e *The flow polynomial*   Let $C^*(\Gamma; u)$ denote the number of nowhere-zero u-flows (see **4k**) on a connected graph Γ with n vertices and m edges. Then

$$C^*(\Gamma; u) = (-1)^m R(\Gamma; -1, -u) = (-1)^{m-n+1} T(\Gamma; 0, 1-u).$$

If Γ is planar and Γ^* is its dual, then (Tutte 1954):

$$C(\Gamma^*; u) = u C^*(\Gamma; u).$$

Thus the problem of finding the flow polynomial of a planar graph is equivalent to finding the chromatic polynomial of its dual. For example, the flow polynomial of a ladder graph can be derived from the chromatic polynomial of its dual, a double pyramid (**9a**).

The general relationship between the flow polynomial of a graph and an 'interaction model' is discussed in Biggs (1977b, Chapter 3).

14f *The flow polynomials of $K_{3,3}$ and O_3.*   From the rank matrix of $K_{3,3}$ (Chapter 10) and the Tutte matrix of O_3 (**13b**) we can obtain the flow polynomials for these (non-planar) graphs:

$$C^*(K_{3,3}; u) = (u-1)(u-2)(u^2 - 6u + 10);$$
$$C^*(O_3; u) = (u-1)(u-2)(u-3)(u-4)(u^2 - 5u + 10).$$

In both cases there is no graph whose chromatic polynomial is uC^*.

14g *Expansions of the flow polynomial*   Jaeger (1991) obtained an expansion of the flow polynomial of a graph Γ of degree 3 imbedded in the plane. Define an *even subgraph* to be a subgraph $\langle C \rangle$ in which every vertex has even degree. Since Γ has degree 3 this means that every component of $\langle C \rangle$ is a cycle, and so each component can be oriented in one of two ways. Associated with every oriented even subgraph $\langle C \rangle$ is a weight $w(C)$, such that

$$C^*(\Gamma; (u + u^{-1})^2) = \sum_C w(C) u^{2\rho(C)},$$

where $\rho(C)$ is a 'rotation number' depending on the relative orientation of the cycles of $\langle C \rangle$ with respect to the plane in which Γ is embedded.

14h *The superiority of the matrix method*   It follows from the result of Jaeger, Vertigan and Welsh (**131**) that computing the chromatic polynomial is, in general, #P-hard. However, there is some interest in comparing methods of computation, even though they are all 'bad' in theoretical terms.

The matrix method (call it Method $\mathcal{A}$), described in our comments on Theorem 14.1, has been used only rarely (Biggs 1973b). However, Anthony (1990) showed that it is more efficient than the method of deletion and contraction (Method $\mathcal{B}$), even when that method incorporates rules for curtailing the computation. Specifically, the worst-case running time of Method $\mathcal{A}$ for a graph with n vertices and m edges is of the order of $\binom{m}{n-1} n^2 m$. If $T_{\mathcal{A}}(n)$ and $T_{\mathcal{B}}(n)$ denote the worst-case running times of the respective methods for any sequence of graphs such that Γ_n has n vertices and the average degree $\Delta(n) \to \infty$ as $n \to \infty$, we have

$$\frac{\log T_{\mathcal{B}}(n)}{\log T_{\mathcal{A}}(n)} \to \infty \quad \text{as } n \to \infty.$$

Symmetry and regularity

15

Automorphisms of graphs

An *automorphism* of a (simple) graph Γ is a permutation π of $V\Gamma$ which has the property that $\{u,v\}$ is an edge of Γ if and only if $\{\pi(u), \pi(v)\}$ is an edge of Γ. The set of all automorphisms of Γ, with the operation of composition, is the *automorphism group* of Γ, denoted by $\mathrm{Aut}(\Gamma)$.

Some basic properties of automorphisms are direct consequences of the definitions. For example, if two vertices x and y belong to the same *orbit*, that is, if there is an automorphism α such that $\alpha(x) = y$, then x and y have the same degree. This, and other similar results, will be taken for granted in our exposition.

We say that Γ is *vertex-transitive* if $\mathrm{Aut}(\Gamma)$ acts transitively on $V\Gamma$, that is, if there is just one orbit. This means that given any two vertices u and v there is an automorphism $\pi \in \mathrm{Aut}(\Gamma)$ such that $\pi(u) = v$. The action of $\mathrm{Aut}(\Gamma)$ on $V\Gamma$ induces an action on $E\Gamma$, by the rule $\pi\{x, y\} = \{\pi(x), \pi(y)\}$, and we say that Γ is *edge-transitive* if this action is transitive, in other words, if given any pair of edges there is an automorphism which transforms one into the other. It is easy to construct graphs which are vertex-transitive but not edge-transitive: the ladder graph L_3 is a simple example. In the opposite direction we have the following result.

Proposition 15.1 *If a connected graph is edge-transitive but not vertex-transitive, then it is bipartite.*

Proof Let $\{x, y\}$ be an edge of Γ, and let X and Y denote the orbits containing x and y respectively under the action of $\mathrm{Aut}(\Gamma)$ on the vertices. It follows from the definition of an orbit that X and Y are either disjoint or identical. Since Γ is connected, every vertex z is in some edge $\{z, w\}$, and since Γ is edge-transitive, z belongs to either X or Y. Thus $X \cup Y = V\Gamma$. If $X = Y = V\Gamma$ then Γ would be vertex-transitive, contrary to hypothesis; consequently $X \cap Y$ is empty. Every edge of Γ has one end in X and one end in Y, so Γ is bipartite. $\qquad\square$

The complete bipartite graph $K_{a,b}$ with $a \neq b$ is an obvious example of a graph which is edge-transitive but not vertex-transitive. In this case the graph is not regular, and it is not vertex-transitive for that reason, because it is clear that in a vertex-transitive graph each vertex must have the same degree. Examples of *regular* graphs which are edge-transitive but not vertex-transitive are not quite so obvious, but examples are known (see **15c**).

The next proposition establishes a link between the spectrum of a graph and its automorphism group. We shall suppose that $V\Gamma$ is the set $\{v_1, v_2, \ldots, v_n\}$, and that the rows and columns of the adjacency matrix of Γ are labelled in the usual way. A permutation π of $V\Gamma$ can be represented by a *permutation matrix* $\mathbf{P} = (p_{ij})$, where $p_{ij} = 1$ if $v_i = \pi(v_j)$, and $p_{ij} = 0$ otherwise.

Proposition 15.2 *Let $\mathbf{A}$ be the adjacency matrix of a graph Γ, and π a permutation of $V\Gamma$. Then π is an automorphism of Γ if and only if $\mathbf{PA} = \mathbf{AP}$, where $\mathbf{P}$ is the permutation matrix representing π.*

Proof Let $v_h = \pi(v_i)$ and $v_k = \pi(v_j)$. Then we have

$$(\mathbf{PA})_{hj} = \Sigma p_{hl} a_{lj} = a_{ij};$$
$$(\mathbf{AP})_{hj} = \Sigma a_{hl} p_{lj} = a_{hk}.$$

Consequently, $\mathbf{AP} = \mathbf{PA}$ if and only if v_i and v_j are adjacent whenever v_h and v_k are adjacent; that is, if and only if π is an automorphism of Γ. $\qquad\square$

A consequence of this result is that, loosely speaking, automorphisms produce multiple eigenvectors corresponding to a given eigenvalue. To be precise, suppose $\mathbf{x}$ is an eigenvector of $\mathbf{A}$ corresponding to the eigenvalue λ. Then we have

$$\mathbf{APx} = \mathbf{PAx} = \mathbf{P}\lambda\mathbf{x} = \lambda\mathbf{Px}.$$

This means that $\mathbf{Px}$ is also an eigenvector of $\mathbf{A}$ corresponding to the eigenvalue λ. If $\mathbf{x}$ and $\mathbf{Px}$ are linearly independent we conclude that λ is not a simple eigenvalue. The following results provide a complete description of what happens when λ is simple.

Lemma 15.3 *Let λ be an simple eigenvalue of Γ, and let $\mathbf{x}$ be a corresponding eigenvector, with real components. If the permutation matrix $\mathbf{P}$ represents an automorphism of Γ then $\mathbf{Px} = \pm\mathbf{x}$.*

Proof If λ has multiplicity one, $\mathbf{x}$ and $\mathbf{Px}$ are linearly dependent; that is $\mathbf{Px} = \mu\mathbf{x}$ for some complex number μ. Since $\mathbf{x}$ and $\mathbf{P}$ are real, μ is real; and, since $\mathbf{P}^s = \mathbf{I}$ for some natural number $s \geq 1$, it follows that μ is an sth root of unity. Consequently $\mu = \pm 1$ and the lemma is proved. $\square$

Theorem 15.4 (Mowshowitz 1969, Petersdorf and Sachs 1969) *If all the eigenvalues of the graph Γ are simple, every automorphism of Γ (apart from the identity) has order 2.*

Proof Suppose that every eigenvalue of Γ has multiplicity one. Then, for any permutation matrix $\mathbf{P}$ representing an automorphism of Γ, and any eigenvector $\mathbf{x}$, we have $\mathbf{P}^2\mathbf{x} = \mathbf{x}$. The space spanned by the eigenvectors is the whole space of column vectors, and so $\mathbf{P}^2 = \mathbf{I}$. $\square$

Theorem 15.4 characterizes the group of a graph which has the maximum number $n = |V\Gamma|$ of distinct eigenvalues: every element of the group is an involution, and so the group is an elementary abelian 2-group. For example, the theta graph $\Theta_{2,2,1}$ (K_4 with one edge deleted) has automorphism group $\mathbb{Z}_2 \times \mathbb{Z}_2$. The characteristic polynomial is

$$\chi(\Theta_{2,2,1}; \lambda) = \lambda(\lambda + 1)(\lambda^2 - \lambda - 4),$$

and so every eigenvalue is simple. On the other hand if we know that a graph has an automorphism of order at least three, then it must have a multiple eigenvalue. In particular, this means that the $2h$ numbers obtained in **3e** as the eigenvalues of the Möbius ladder M_{2h} cannot all be distinct.

The question of which groups can be the automorphism group of some graph was answered by Frucht (1938). He showed that, for every abstract finite group G, there is a graph Γ whose automorphism group is isomorphic to G. He also proved that the same result holds with Γ restricted to be a regular graph of degree 3 (Frucht 1949). Although there are some gaps in the original proof, satisfactory proofs of the result are now available. For an overview of this subject the reader is referred to Babai (1981). He describes how Frucht's work stimulated a great deal of research, and how it has been extended by several authors to show that the conclusion remains true, even if we specify in advance that Γ must satisfy a number of graph-theoretical conditions.

If we strengthen the question by asking whether every group of permutations of a set X is the automorphism group of some graph with vertex-set X, then the answer is negative. For example, the cyclic permutation-group of order 3 is not the automorphism group of any graph with three vertices. (It is, of course, a *subgroup* of the group of K_3.) This tends to confirm our intuitive impression that there must be some constraints upon the possible symmetry of graphs. One such constraint is the following. If Γ is a connected graph and $\partial(u, v)$ denotes the distance in Γ between the vertices u and v then, for any automorphism α we have

$$\partial(u, v) = \partial(\alpha(u), \alpha(v)).$$

Thus there can be no automorphism which transforms a pair of vertices at distance r into a pair at distance $s \neq r$. The following definition frames conditions which are, in a sense, partially converse to this result.

Definition 15.5 Let Γ be a graph with automorphism group $\mathrm{Aut}(\Gamma)$. We say that Γ is *symmetric* if, for all vertices u,v,x,y of Γ such that u and v are adjacent, and x and y are adjacent, there is an automorphism α in $\mathrm{Aut}(\Gamma)$ for which $\alpha(u) = x$ and $\alpha(v) = y$. We say that Γ is *distance-transitive* if, for all vertices u,v,x,y of Γ such that $\partial(u, v) = \partial(x, y)$, there is an automorphism α in $\mathrm{Aut}(\Gamma)$ satisfying $\alpha(u) = x$ and $\alpha(v) = y$.

It is clear that we have a hierarchy of conditions:

$$\text{distance-transitive} \Rightarrow \text{symmetric} \Rightarrow \text{vertex-transitive}.$$

In the following chapters we shall investigate these conditions in turn, beginning with the weakest one.

Additional Results

15a *How large can an automorphism group be?* For any value of n, the automorphism group of the complete graph K_n contains all the $n!$ permutations of its n vertices: it is the *symmetric group* S_n. Any other graph on n vertices has an automorphism group which is a subgroup of S_n. Since the complete graph is the only connected graph in which each pair of distinct vertices is at the same distance, it is the only connected graph for which the automorphism group can act *doubly-transitively* on the vertex-set.

15b *How small can an automorphism group be?* Except for very small values of n, it is easy to construct a graph with n vertices which has the trivial automorphism group, containing only the identity permutation. For $n \geq 7$ the tree with n vertices shown in Figure 7 is an example.

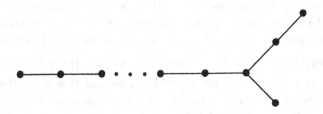

Figure 7: a tree with no non-trivial automorphisms.

In fact, almost all graphs have the trivial automorphism group. The full story is described by Bollobás (1985, Chapter 9).

15c *A regular graph which is edge-transitive but not vertex-transitive* Consider a cube divided into 27 equal cubes, in the manner of Rubik's cube, and let us say that a *row* is a set of three cubes in a row parallel to a side of the big cube. Define a graph whose vertices are the 27 cubes and the 27 rows, a cube-vertex being adjacent to the three row-vertices to which it belongs. This example of a regular edge-transitive graph which is not vertex-transitive is the first of a family of examples due to Bouwer (1972).

15d *The automorphism groups of trees* (Jordan 1869) Let T be a finite tree. Then either (i) T has a vertex v, known as the *centroid*, which is fixed by every automorphism of T, or (ii) T has an edge $\{x, y\}$, known as the *bicentroid*, which is fixed (setwise) by every automorphism of T.

15e *The graphs $P(h, t)$* The 'generalized Petersen graph' $P(h, t)$ is a 3-regular graph with $2h$ vertices $x_0, x_1, \ldots, x_{h-1}, y_0, y_1, \ldots, y_{h-1}$ and edges $\{x_i, y_i\}, \{x_i, x_{i+1}\}, \{y_i, y_{i+t}\}$, for all $i \in \{0, 1, \ldots, h-1\}$, where the subscripts are reduced modulo h. For example, $P(h, 1)$ is the ladder graph L_h, and $P(5, 2)$ is Petersen's graph. Frucht, Graver and Watkins (1971) showed that:
(a) $P(h, t)$ is vertex-transitive if and only if $t^2 \equiv \pm 1 \pmod{h}$, or $(h, t) = (10, 2)$.
(b) $P(h, t)$ is symmetric if and only if (h, t) is one of (4,1), (5,2), (8,3), (10,2), (10,3), (12,5), (24,5).

Case-by-case checking of the latter result shows that $P(h, t)$ is distance-transitive if and only if (h, t) is one of (4,1), (5,2), (10,3).

15f *The connection between* $\text{Aut}(\Gamma)$ *and* $\text{Aut}(L(\Gamma))$ (Whitney 1932c) The automorphism groups of Γ and its line graph $L(\Gamma)$ are not necessarily isomorphic: for example $K_1 = L(K_2)$, so in this case the first group is trivial but the second is not. However, this is a rare phenomenon. There is a group homomorphism $\theta : \text{Aut}(\Gamma) \to \text{Aut}(L(\Gamma))$ defined by

$$(\theta g)\{u, v\} = \{\theta u, \theta v\} \quad \text{where} \quad g \in \text{Aut}(\Gamma), \ \{u, v\} \in E(\Gamma),$$

and we have: (i) θ is a monomorphism provided $\Gamma \neq K_2$; (ii) θ is an epimorphism provided Γ is not K_4, K_4 with one edge deleted, or K_4 with two adjacent edges deleted.

15g *Homogeneous graphs*   A graph Γ is said to be *weakly homogeneous* if, whenever two subsets U_1, U_2 of $V\Gamma$ are such that $\langle U_1 \rangle$ and $\langle U_2 \rangle$ are isomorphic, then at least one isomorphism between them extends to an automorphism of Γ. The complete list of weakly homogeneous graphs is as follows.

(a) The cycle graph C_5.

(b) The disjoint union of $t \geq 1$ copies of the complete graph K_n.

(c) The complete multipartite graphs $K_{s,s,\dots,s}$ with $t \geq 2$ parts of equal size s.

(d) The line graph $L(K_{3,3})$.

A graph is *homogeneous* if, whenever two subsets U_1, U_2 of $V\Gamma$ are such that $\langle U_1 \rangle$ and $\langle U_2 \rangle$ are isomorphic, then *every* isomorphism between them extends to an automorphism of Γ. It is obvious that a homogeneous graph is weakly homogeneous, and somewhat surprisingly the converse is also true. This result has a contorted history. The 1974 version of this book caused some confusion by attributing to Sheehan the classification of weakly homogeneous graphs given above. In fact, Sheehan (1974) obtained the classification of homogeneous graphs. Gardiner observed the error in the book, and then (1976) obtained the same list for the weakly homogeneous case by an independent method. Finally Ronse (1978) showed directly that a weakly homogeneous graph is homogeneous.

15h *Graphs which are transitive on vertices and edges*   Let Γ be a graph for which $\text{Aut}(\Gamma)$ acts transitively on both vertices and edges. Then Γ is a regular graph, and if its degree is odd, it is symmetric (Tutte 1966). If its degree is even, the conclusion may be false, as was first shown by Bouwer (1970). Holt (1981) gave an example of a 4-regular graph with 27 vertices which is vertex-transitive and edge-transitive, but not symmetric, and Alspach, Marušič and Nowitz (1993) showed that Holt's example is the smallest possible.

15i *Graphs with a given group* (Izbicki 1960)   Let an abstract finite group G and natural numbers r and s satisfying $r \geq 3$, $2 \leq s \leq r$ be given. Then there are infinitely many graphs Γ with the properties:

    (a) Aut(Γ) is isomorphic to G;

    (b) Γ is regular of degree r;

    (c) the chromatic number of Γ is s.

16

Vertex-transitive graphs

In this chapter we study graphs Γ for which the automorphism group acts transitively on $V\Gamma$. As we have already noted in the previous chapter, vertex-transitivity implies that every vertex has the same degree, so Γ is a regular graph.

We shall use the following standard results on transitive permutation groups. Let $G = \mathrm{Aut}(\Gamma)$ and let G_v denote the *stabilizer* subgroup for the vertex v; that is, the subgroup of G containing those automorphisms which fix v. In the vertex-transitive case all stabilizer subgroups G_v ($v \in V\Gamma$) are conjugate in G, and consequently isomorphic. The index of G_v in G is given by the equation

$$|G : G_v| = |G|/|G_v| = |V\Gamma|.$$

If each stabilizer G_v is the identity group, then every element of G (except the identity) does not fix any vertex, and we say that G acts *regularly* on $V\Gamma$. In this case the order of G is equal to the number of vertices.

There is a standard construction, due originally to Cayley (1878), which enables us to construct many, but not all, vertex-transitive graphs. We shall give a streamlined version which has proved to be well-adapted to the needs of algebraic graph theory. Let G be any abstract finite group, with identity 1; and suppose Ω is a set of generators for G, with the properties:

(i) $x \in \Omega \Rightarrow x^{-1} \in \Omega$;   (ii) $1 \notin \Omega$.

Definition 16.1 The *Cayley graph* $\Gamma = \Gamma(G, \Omega)$ is the simple graph whose vertex-set and edge-set are defined as follows:

$$V\Gamma = G; \quad E\Gamma = \{\{g, h\} \mid g^{-1}h \in \Omega\}.$$

Simple verifications show that $E\Gamma$ is well-defined, and that $\Gamma(G, \Omega)$ is a connected graph. For example, if G is the symmetric group S_3, and $\Omega = \{(12), (23), (13)\}$, then the Cayley graph $\Gamma(G, \Omega)$ is isomorphic to $K_{3,3}$ (Figure 8).

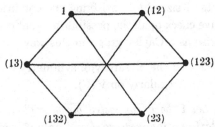

Figure 8: $K_{3,3}$ as a Cayley graph for S_3.

Proposition 16.2 (1) *The Cayley graph $\Gamma(G, \Omega)$ is vertex-transitive.* (2) *Suppose that π is an automorphism of the group G such that $\pi(\Omega) = \Omega$. Then π, regarded as a permutation of the vertices of $\Gamma(G, \Omega)$, is a graph automorphism fixing the vertex 1.*

Proof (1) For each g in G we may define a permutation $\bar{g}$ of $V\Gamma = G$ by the rule $\bar{g}(h) = gh$ ($h \in G$). This permutation is an automorphism of Γ, for

$$\{h, k\} \in E\Gamma \Rightarrow h^{-1}k \in \Omega$$
$$\Rightarrow (gh)^{-1}gk \in \Omega$$
$$\Rightarrow \{\bar{g}(h), \bar{g}(k)\} \in E\Gamma.$$

The set of all $\bar{g}$ ($g \in G$) constitutes a group $\overline{G}$ (isomorphic with G), which is a subgroup of the full group of automorphisms of $\Gamma(G, \Omega)$, and acts transitively on the vertices.

(2) Since π is a group automorphism it must fix the vertex 1. Furthermore, π is a graph automorphism, since

$$\{h, k\} \in E\Gamma \Rightarrow h^{-1}k \in \Omega \Rightarrow \pi(h^{-1}k) \in \Omega$$
$$\Rightarrow \pi(h)^{-1}\pi(k) \in \Omega$$
$$\Rightarrow \{\pi(h), \pi(k)\} \in E\Gamma. \qquad \square$$

The second part of this proposition implies that the automorphism group of a Cayley graph $\Gamma(G, \Omega)$ will often be strictly larger than $\overline{G}$.

In the example illustrated in Figure 8, every group automorphism of S_3 fixes Ω setwise, and so it follows that the stabilizer of the vertex 1 has order at least 6. In fact, the order of the stabilizer is 12, and $|\mathrm{Aut}(K_{3,3})| = 72$.

Not every vertex-transitive graph is a Cayley graph; for example Petersen's graph O_3 is not a Cayley graph. This statement can be checked by noting that there are only two groups of order 10 and they have few generating sets of size three satisfying the conditions in Definition 16.1. An exhaustive check of all the possibilities confirms that Petersen's graph does not arise as a Cayley graph in this way.

We begin our study of the hierarchy of symmetry conditions with the case when $\mathrm{Aut}(\Gamma)$ acts regularly on $V(\Gamma)$.

Lemma 16.3 *Let Γ be a connected graph. Then a subgroup H of $\mathrm{Aut}(\Gamma)$ acts regularly on the vertices if and only if Γ is isomorphic to a Cayley graph $\Gamma(H, \Omega)$, for some set Ω which generates H.*

Proof Suppose $V\Gamma = \{v_1, v_2, \ldots, v_n\}$, and H is a subgroup of $\mathrm{Aut}(\Gamma)$ acting regularly on $V\Gamma$. Then, for $1 \le i \le n$ there is a unique $h_i \in H$ such that $h_i(v_1) = v_i$. Let

$$\Omega = \{h_i \in H \mid v_i \text{ is adjacent to } v_1 \text{ in } \Gamma\}.$$

Simple checks show that Ω satisfies the two conditions required by Definition 16.1 and that the bijection $v_i \leftrightarrow h_i$ is a graph isomorphism of Γ with $\Gamma(H, \Omega)$. Conversely if $\Gamma = \Gamma(H, \Omega)$ then the group $\overline{H}$ defined in the proof of Proposition 16.2 acts regularly on $V\Gamma$, and $\overline{H} \approx H$. □

Lemma 16.3 shows that if $\mathrm{Aut}(\Gamma)$ itself acts regularly on $V\Gamma$, then Γ is a Cayley graph $\Gamma(\mathrm{Aut}(\Gamma), \Omega)$.

Definition 16.4 A finite abstract group G admits a *graphical regular representation*, or GRR, if there is a graph Γ such that G is isomorphic with $\mathrm{Aut}(\Gamma)$, and $\mathrm{Aut}(\Gamma)$ acts regularly on $V\Gamma$.

The question of which abstract groups admit a GRR was answered completely in the late 1970's (see **16g**). It turns out that the second part of Proposition 16.2 is essentially the only obstacle to there being a GRR for G. In other words, a group G has no GRR if and only if every generating set Ω for G which satisfies conditions (i) and (ii) is such that there is an automorphism of G fixing Ω setwise.

As an example of the ideas involved we show that the group S_3 admits no graphical regular representation. If there were a suitable graph Γ, then it would be a Cayley graph $\Gamma(S_3, \Omega)$. Now, it is easy to check by an exhaustive search that, for any generating set Ω satisfying conditions (i) and (ii) on p. 122, there is some automorphism of S_3 fixing Ω setwise. Thus, by part (2) of Proposition 16.2, the automorphism group of a Cayley graph $\Gamma(S_3, \Omega)$ is strictly larger than S_3.

In the case of transitive abelian groups, precise information is provided by the next proposition.

Proposition 16.5 *Let Γ be a vertex-transitive graph whose automorphism group $G = \mathrm{Aut}(\Gamma)$ is abelian. Then G acts regularly on $V\Gamma$, and G is an elementary abelian 2-group.*

Proof If g and h are elements of the abelian group G, and g fixes v, then $gh(v) = hg(v) = h(v)$ so that g fixes $h(v)$ also. If G is transitive, every vertex is of the form $h(v)$ for some h in G, so g fixes every vertex. That is, $g = 1$.

Thus G acts regularly on $V\Gamma$ and so, by Lemma 16.3, Γ is a Cayley graph $\Gamma(G, \Omega)$. Now since G is Abelian, the function $g \mapsto g^{-1}$ is an automorphism of G, and it fixes Ω setwise. If this automorphism were non-trivial, then part (2) of Proposition 16.2 would imply that G is not regular. Thus $g = g^{-1}$ for all $g \in G$, and every element of G has order 2. $\quad\square$

We now turn to a discussion of some simple spectral properties of vertex-transitive graphs. A vertex-transitive graph Γ is necessarily a regular graph, and so its spectrum has the properties which are stated in Proposition 3.1. In particular, if Γ is connected and regular of degree k, then k is a simple eigenvalue of Γ. It turns out that we can use the vertex-transitivity property to characterize the simple eigenvalues of Γ.

Proposition 16.6 (Petersdorf and Sachs 1969)   *Let Γ be a vertex-transitive graph which has degree k, and let λ be a simple eigenvalue of Γ. If $|V\Gamma|$ is odd, then $\lambda = k$. If $|V\Gamma|$ is even, then λ is one of the integers $2\alpha - k$ ($0 \le \alpha \le k$).*

Proof Let $\mathbf{x}$ be a real eigenvector corresponding to the simple eigenvalue λ, and let $\mathbf{P}$ be a permutation matrix representing an automorphism π of Γ. If $\pi(v_i) = v_j$, then, by Lemma 15.3,

$$x_i = (\mathbf{P}\mathbf{x})_j = \pm x_j.$$

Since Γ is vertex-transitive, we deduce that all the entries of $\mathbf{x}$ have the same absolute value. Now, since $\mathbf{u} = [1, 1 \ldots, 1]^t$ is an eigenvector corresponding to the eigenvalue k, if $\lambda \ne k$ we must have $\mathbf{u}^t\mathbf{x} = 0$, that is $\sum x_i = 0$. This is impossible for an odd number of summands of equal absolute value, and so our first statement is proved.

If Γ has an even number of vertices, choose a vertex v_i of Γ and suppose that, of the vertices v_j adjacent to v_i, a number α have $x_j = x_i$ while $k - \alpha$ have $x_j = -x_i$. Since $(\mathbf{A}x)_i = \lambda x_i$ it follows that $\sum' x_j = \lambda x_i$, where the sum is taken over vertices adjacent to v_i. Thus

$$\alpha x_i - (k - \alpha)x_i = \lambda x_i.$$

whence $\lambda = 2\alpha - k$. $\square$

For example, the only numbers which can be simple eigenvalues of a 3-regular vertex-transitive graph are $3, 1, -1, -3$. This statement is false if we assume merely that the graph is regular of degree 3: many examples can be found in [CvDS, pp. 292–305].

If we strengthen the assumptions by postulating that Γ is symmetric, then the simple eigenvalues are restricted still further.

Proposition 16.7   *Let Γ be a symmetric graph of degree k, and let λ be a simple eigenvalue of Γ. Then $\lambda = \pm k$.*

Proof   We continue to use the notation of the previous proof. Let v_j and v_l be any two vertices adjacent to v_i; then there is an automorphism π of Γ such that $\pi(v_i) = v_i$ and $\pi(v_j) = v_l$. If $\mathbf{P}$ is the permutation matrix representing π, then $\pi(v_i) = v_i$ implies that $\mathbf{P}x = \mathbf{x}$ and so $x_j = x_l$. Thus $\alpha = 0$ or k, and $\lambda = \pm k$.

We remark that the eigenvalue $-k$ occurs, and is necessarily simple, if and only if Γ is bipartite.

Additional Results

16a *Circulant graphs*   A circulant graph is vertex-transitive, and a connected circulant graph is a Cayley graph $\Gamma(\mathbf{Z}_n, \Omega)$ for a cyclic group $\mathbf{Z}_n$. Ádám (1967) conjectured that if two such graphs $\Gamma(\mathbf{Z}_n, \Omega)$ and $\Gamma(\mathbf{Z}_n, \Omega')$ are isomorphic then $\Omega' = z\Omega$ for some invertible element z in $\mathbf{Z}_n$. Elspas and Turner (1970) showed that the conjecture is true if n is a prime, or if the graphs have only simple eigenvalues, but false in general. Parsons (1980) showed that it is true if both graphs have vertex-neighbourhoods isomorphic to the cycle C_k.

16b *The ladder graphs as Cayley graphs*   The *dihedral group* D_{2n} of order $2n$ is defined by the presentation

$$D_{2n} = \langle x, y \mid x^n = y^2 = (xy)^2 = 1 \rangle.$$

The Cayley graph of D_{2n} with respect to the generating set $\{x, x^{-1}, y\}$ is the ladder graph L_n.

16c *Cayley graphs for the tetrahedral and icosahedral groups* The alternating group A_n is the subgroup of index two in S_n containing all the even permutations. The groups A_4 and A_5 are sometimes known as the *tetrahedral* and *icosahedral* groups because they are isomorphic with groups of rotations of the respective polyhedra. Both groups can be represented by planar Cayley graphs. A Cayley graph for A_4 is shown in Figure 9.

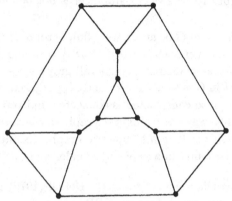

Figure 9: a Cayley graph for A_4.

A Cayley graph for A_5 is the skeleton of the famous carbon-60 structure, also known as buckminsterfullerene, or the 'buckie-ball', or the 'soccer ball'.

16d *The stabilizer of a vertex-neighbourhood* Suppose that Γ is a vertex-transitive graph with $G = \text{Aut}(\Gamma)$. For any vertex v of Γ, define

$$L_v = \{g \in G_v \mid g \text{ fixes each vertex adjacent to } v\}.$$

Then L_v is a normal subgroup of G_v. More explicitly, there is a homomorphism from G_v into the group of all permutations of the neighbours of v, with kernel L_v. It follows from this that $|G_v : L_v| \leq k!$, where k is the degree.

16e *The order of the vertex-stabilizer* Let H_n be the graph formed by linking together n 'units' of the form shown in Figure 10 so that they form a complete circuit. Then the graphs H_n are vertex-transitive and the order of the vertex-stabilizer (2^n) is not bounded in terms of the degree.
On the other hand, in a symmetric graph the order of the vertex-stabilizer is bounded in terms of the degree. See **17g**.

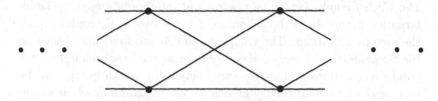

Figure 10: the vertex-stabilizer is not bounded.

16f *Coset graphs*   Let G be an abstract finite group, H a subgroup of G, and Ω a subset of $G \setminus H$ such that $1 \notin \Omega$, $\Omega^{-1} = \Omega$, and $H \cup \Omega$ generates G. The simplest way of defining a (general) graph whose vertices are the right cosets of H in G is to make Hg_1 and Hg_2 adjacent whenever $g_2 g_1^{-1}$ is in Ω. The graph so constructed is connected and vertex-transitive.

There are other ways of defining a graph whose vertices are cosets, and some of them result in a symmetric graph. Examples and further references may be found in a paper by Conder and Lorimer (1989).

16g *Graphical regular representations*   Hetzel (1976) proved that the only solvable groups which have no GRR are:
(a) abelian groups of exponent greater than 2;
(b) generalized dicyclic groups;
(c) thirteen 'exceptional' groups, such as the elementary abelian groups $\mathbb{Z}_2^2, \mathbb{Z}_2^3, \mathbb{Z}_2^4$, the dihedral groups D_6, D_8, D_{10}, and the alternating group A_4.

This work subsumed earlier results by several other authors. Godsil (1981) showed that every non-solvable group has a GRR, so the list given above is the complete list of groups which have no GRR.

16h *The eigenvalues of a Cayley graph* (Babai 1979)   Let $\Gamma(G, \Omega)$ be a Cayley graph and suppose that the irreducible characters of G are $\chi_1, \chi_2, \ldots, \chi_c$, with degrees $n_1, n_2, \ldots, n_c$ respectively. Then the eigenvalues of Γ fall into families $(\Lambda_i)_j$, $1 \leq i \leq c$, $1 \leq j \leq n_i$ such that each $(\Lambda_i)_j$ contains n_i eigenvalues, all with a common value λ_{ij}. (Note that the total number of eigenvalues is thus $\sum n_i^2$, which is the correct number $|G|$.) The sum of the tth powers of the λ_{ij} corresponding to a given character χ_i satisfies

$$\sum_{j=1}^{n_i} \lambda_{ij}^t = \sum \chi_i(\omega_1 \omega_2 \ldots \omega_t),$$

where the sum on the right-hand side is taken over all products of t elements of Ω.

16i *The Paley graphs*   Denote the additive group of the field $GF(q)$ by G_q and let Ω be the set of non-zero squares in $GF(q)$. If $q \equiv 1$ (mod 4) then Ω generates G_q and satisfies the conditions at the foot of p. 122 (remembering that the identity of G_q is the zero element of the field). The *Paley graph* $P(q)$ is the Cayley graph $\Gamma(G_q, \Omega)$. These graphs are strongly regular and self-complementary. If q is the rth power of a prime, the order of $\text{Aut}(P(q))$ is $rq(q-1)/2$.

16j *Graphs with a specified vertex-neighbourhood*   A graph is said to be *locally* K if, for each vertex v, the subgraph induced by the neighbours of v is isomorphic to K. For example, the graphs which are locally Petersen were determined by Hall (1980): there are just three of them, having 21, 63, and 65 vertices. Many other papers on this topic are listed by Blokhuis and Brouwer (1992).

16k *Generators for the automorphism group*   Let Γ be a connected vertex-transitive graph and let G_v denote the stabilizer of the vertex v. If h is any automorphism of Γ for which $\partial(v, h(v)) = 1$, and Γ is symmetric, then h and G_v generate $\text{Aut}(\Gamma)$.

17

Symmetric graphs

The condition of vertex-transitivity is not a very powerful one, as is demonstrated by the fact that we can construct at least one vertex-transitive graph from each finite group, by means of the Cayley graph construction. A vertex-transitive graph is symmetric if and only if each vertex-stabilizer G_v acts transitively on the set of vertices adjacent to v. For example, there are just two distinct 3-regular graphs with 6 vertices; one is $K_{3,3}$ and the other is the ladder L_3. Both these graphs are vertex-transitive, and $K_{3,3}$ is symmetric, but L_3 is not because there are two 'kinds' of edges at each vertex.

Although the property of being symmetric is apparently only slightly stronger than vertex-transitivity, symmetric graphs do have distinctive properties which are not shared by all vertex-transitive graphs. This was first demonstrated by Tutte (1947a) in the case of 3-regular graphs. More recently his results have been extended to graphs of higher degree, and it has become apparent that the results are closely related to fundamental classification theorems in group theory. (See **17a, 17f, 17g.**)

We begin by defining a *t-arc* $[\alpha]$ in a graph Γ to be a sequence $(\alpha_0, \alpha_1, \ldots, \alpha_t)$ of $t+1$ vertices of Γ, with the properties that $\{\alpha_{i-1}, \alpha_i\}$ is in $E\Gamma$ for $1 \leq i \leq t$, and $\alpha_{i-1} \neq \alpha_{i+1}$ for $1 \leq i \leq t-1$. A t-arc is not quite the same thing as the sequence of vertices underlying a path of length t, because it is convenient to allow repeated vertices. We regard a single vertex v as a 0-arc $[v]$. If $\beta = (\beta_0, \beta_1, \ldots, \beta_s)$ is an s-arc in Γ, then we write $[\alpha.\beta]$ for the sequence $(\alpha_0, \ldots, \alpha_t, \beta_0, \ldots, \beta_s)$, provided that this is a $(t + s + 1)$-arc; that is, provided α_t is adjacent to β_0 and $\alpha_{t-1} \neq \beta_0, \alpha_t \neq \beta_1$.

Definition 17.1   A graph Γ is *t-transitive* $(t \geq 1)$ if its automorphism group is transitive on the set of t-arcs in Γ, but not transitive on the set of $(t + 1)$-arcs in Γ.

There is little risk of confusion with the concept of multiple transitivity used in the general theory of permutation groups, since (as was noted in **15a**) the only graphs which are multiply transitive in that sense are the complete graphs. We observe that the automorphism group is transitive on 1-arcs if and only if Γ is symmetric (since a 1-arc is just a pair of adjacent vertices). Consequently, any symmetric graph is t-transitive for some $t \geq 1$.

The only connected graph of degree one is K_2, and this graph is 1-transitive. The only connected graphs of degree two are the cycle graphs C_n $(n \geq 3)$, and these are anomalous in that they are transitive on t-arcs for all $t \geq 1$. From now on, we shall usually assume that the graphs under consideration are connected and regular of degree not less than three. For such graphs we have the following elementary inequality.

Proposition 17.2   *Let Γ be a t-transitive graph whose degree is at least three and whose girth is g. Then*

$$t \leq \frac{1}{2}(g + 2).$$

Proof Γ contains a cycle of length g, which is, in particular, a g-arc. Because the degree is at least three, we can alter one edge of this g-arc to obtain a g-arc whose ends do not coincide. Clearly no automorphism of Γ can take a g-arc of the first kind to a g-arc of the second kind; so it follows that $t < g$.

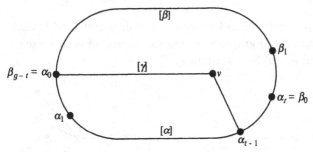

Figure 11: illustrating the proof of Proposition 17.2.

Consequently, if we select a cycle of length g in Γ, then there is a t-arc

$[\alpha]$, without repeated vertices, contained in it. Let $[\beta]$ be the $(g-t)$-arc beginning at α_t and ending at α_0 which completes the cycle of length g. Also let v be a vertex adjacent to α_{t-1}, but which is not α_{t-2} or α_t; this situation is depicted in Figure 11. Since Γ is t-transitive, there is an automorphism taking the t-arc $[\alpha]$ to the t-arc $(\alpha_0, \alpha_1, \ldots, \alpha_{t-1}, v)$. This automorphism must take the $(g-t+1)$-arc $[\alpha_{t-1}.\beta]$ to another $(g-t+1)$-arc $[\alpha_{t-1}.\gamma]$, where $\gamma_0 = v$ and $\gamma_{g-t} = \alpha_0$. The two arcs $[\alpha_{t-1}.\beta]$ and $[\alpha_{t-1}.\gamma]$ may overlap, but they define a cycle of length at most $2(g-t+1)$. Hence $g \le 2(g-t+1)$, that is, $g \ge 2t-2$. $\quad\square$

Definition 17.3 Let $[\alpha]$ and $[\beta]$ be any two s-arcs in a graph Γ. We say that $[\beta]$ is a *successor* of $[\alpha]$ if $\beta_i = \alpha_{i+1}$ $(0 \le i \le s-1)$.

It is helpful to think of the operation of taking a successor of $[\alpha]$ in terms of 'shunting' $[\alpha]$ through one step in Γ. Suppose we ask whether repeated shunting will transform a given s-arc into any other. If there are vertices of degree one in Γ then our shunting might be halted in a 'siding', while if all vertices have degree two we cannot reverse the direction of our 'train'. However, if each vertex of Γ has degree not less than three, and Γ is connected, then our intuition is correct and the shunting procedure always works. The proof of this requires careful examination of several cases, and may be found in Tutte's book (Tutte 1966, pp. 56–58). Formally the result is as follows.

Lemma 17.4 *Let Γ be a connected graph in which the degree of each vertex is at least three. If $s \ge 1$ and $[\alpha], [\beta]$ are any two s-arcs in Γ, then there is a finite sequence $[\alpha^{(i)}]$ $(1 \le i \le l)$ of s-arcs in Γ such that $[\alpha^{(1)}] = [\alpha], [\alpha^{(l)}] = [\beta]$, and $[\alpha^{(i+1)}]$ is a successor of $[\alpha^{(i)}]$ for $1 \le i \le l-1$.* $\quad\square$

We can now state and prove a convenient test for t-transitivity. Let Γ be a connected graph in which the degree of each vertex is at least three, and let $[\alpha]$ be a t-arc in Γ.

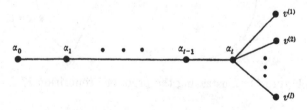

Figure 12: a t-arc and its successors.

Suppose (as in Figure 12) that the vertices adjacent to α_t are α_{t-1} and $v^{(1)}, v^{(2)}, \ldots, v^{(l)}$, and let $[\beta^{(i)}]$ denote the t-arc $(\alpha_1, \alpha_2, \ldots, \alpha_t, v^{(i)})$ for $1 \leq i \leq l$, so that each $[\beta^{(i)}]$ is a successor of $[\alpha]$.

Theorem 17.5 *Let Γ be a connected k-regular graph with $l = k - 1 \geq 3$, and let $[\alpha]$ be a t-arc in Γ. Then $\mathrm{Aut}(\Gamma)$ is transitive on t-arcs if and only if it contains automorphisms $g_1, g_2, \ldots, g_l$ such that $g_i[\alpha] = [\beta^{(i)}]\ (1 \leq i \leq l)$.*

Proof The condition is clearly satisfied if $\mathrm{Aut}(\Gamma)$ is transitive on t-arcs. Conversely, suppose the relevant automorphisms $g_1, g_2, \ldots, g_l$ can be found; then they generate a subgroup $H = \langle g_1, g_2, \ldots, g_l \rangle$ of $\mathrm{Aut}(\Gamma)$, and we shall show that H is transitive on t-arcs.

Let $[\theta]$ be a t-arc in the orbit of $[\alpha]$ under H; thus $[\theta] = h[\alpha]$ for some $h \in H$. If $[\phi]$ is any successor of $[\theta]$, then $h^{-1}[\phi]$ is a successor of $[\alpha]$, and so $[\phi] = hg_i[\alpha]$ for some $i \in \{1, 2, \ldots, l\}$. That is, $[\phi]$ is also in the orbit of $[\alpha]$ under H. Now Lemma 17.4 tells us that all t-arcs can be obtained from $[\alpha]$ by repeatedly taking successors, and so all t-arcs are in the orbit of $[\alpha]$ under H. $\qquad\square$

As an example, consider Petersen's graph O_3, whose vertices are the unordered pairs from the set $\{1, 2, 3, 4, 5\}$ with disjoint pairs being adjacent. The automorphism group is the group of all permutations of $\{1, 2, 3, 4, 5\}$, acting in the obvious way on the vertices. Since the girth of O_3 is 5, Proposition 17.2 tells us that the graph is at most 3-transitive. The 3-arc $[\alpha] = (12, 34, 15, 23)$ has two successors: $[\beta^{(1)}] = (34, 15, 23, 14)$ and $[\beta^{(2)}] = (34, 15, 23, 45)$. The automorphism $(13)(245)$ takes $[\alpha]$ to $[\beta^{(1)}]$ and the automorphism (13524) takes $[\alpha]$ to $[\beta^{(2)}]$, hence O_3 is 3-transitive.

In addition to its usefulness as a test for t-transitivity, Theorem 17.5 also provides a starting point for theoretical investigations into the structure of t-transitive graphs. Suppose that Γ is a connected t-transitive graph ($t \geq 1$), which is regular of degree $k \geq 3$, and let $[\alpha]$ be a given t-arc in Γ.

Definition 17.6 The *stabilizer sequence* of $[\alpha]$ is the sequence

$$\mathrm{Aut}(\Gamma) = G > F_t > F_{t-1} > \ldots > F_1 > F_0$$

of subgroups of $\mathrm{Aut}(\Gamma)$, where $F_i\ (0 \leq i \leq t)$ is defined to be the pointwise stabilizer of the set $\{\alpha_0, \alpha_1, \ldots, \alpha_{t-i}\}$.

In the case of Petersen's graph, with respect to the 3-arc $(12, 34, 15, 23)$, the group F_0 is trivial, F_1 is the group of order 2 generated by (34), F_2

is the group of order 4 generated by (34) and (12), and F_3 is the group of order 12 generated by (34), (12) and (345).

In general, since G is transitive on s-arcs ($1 \leq s \leq t$), all stabilizer sequences of t-arcs are conjugate in G, and consequently we shall often omit explicit reference to $[\alpha]$.

The order of each group occuring in the stabilizer sequence is determined by the order of F_0, as follows. Since F_t is the stabilizer of the single vertex α_0 in the vertex-transitive group G, it follows that $|G : F_t| = n = |V\Gamma|$. Since G is transitive on 1-arcs, F_t acts transitively on the k vertices adjacent to α_0 and F_{t-1} is the stabilizer of the vertex α_1 in this action; consequently $|F_t : F_{t-1}| = k$. Since G is transitive on s-arcs ($2 \leq s \leq t$), the group F_{t-s+1} acts transitively on the $k-1$ vertices adjacent to α_{s-1} (other than α_{s-2}), and F_{t-s} is the stabilizer of the vertex α_s in this action; consequently $|F_{t-s+1} : F_{t-s}| = k - 1$ for $2 \leq s \leq t$.

Thus we have

$$|F_s| = (k-1)^s|F_0| \qquad (0 \leq s \leq t-1),$$
$$|F_t| = k(k-1)^{t-1}|F_0|,$$
$$|G| = nk(k-1)^{t-1}|F_0|.$$

This confirms our earlier observations about Petersen's graph, where we have $t = 3$ and $|F_0| = 1$, so that $|F_1| = 2$, $|F_2| = 4$, $|F_3| = 12$ and $|G| = 120$.

We shall now explain how the properties of the stabilizer sequence can be conveniently discussed in terms of the set $\{g_1, g_2, \ldots, g_l\}$ of $l = k - 1$ automorphisms whose existence is guaranteed by Theorem 17.5. Define an increasing sequence of subsets of $G = \text{Aut}(\Gamma)$, denoted by $\{1\} = Y_0 \subseteq Y_1 \subseteq Y_2 \subseteq \ldots$, as follows:

$$Y_i = \{g_a^{-j}g_b^j \mid a, b \in \{1, 2, \ldots, l\} \text{ and } 1 \leq j \leq i\}.$$

Proposition 17.7   (1) *If* $1 \leq i \leq t$, *then* Y_i *is a subset of* F_i, *but not a subset of* F_{i-1}. (2) *If* $0 \leq i \leq t$, *then* F_i *is the subgroup of* G *generated by* Y_i *and* F_0.

Proof   (1) For $1 \leq a \leq l$, we have $g_a^r(\alpha_j) = \alpha_{j+r}$ provided that both j and $j + r$ lie between 0 and t. Also, $g_a^{t-j+1}(\alpha_j) = v^{(a)}$. It follows that $g_a^{-j}g_b^j$ fixes $\alpha_0, \alpha_1, \ldots, \alpha_{t-i}$ for all $j \leq i$, and so $Y_i \subseteq F_i$. If it were true that $Y_i \subseteq F_{i-1}$, then $g_a^{-i}g_b^i$ would fix α_{t-i+1}, but this means that $g_a^i(\alpha_{t-i+1}) = g_b^i(\alpha_{t-i+1})$, that is $v^{(a)} = v^{(b)}$. Since this is false for $a \neq b$, we have $Y_i \not\subseteq F_{i-1}$.

(2) Suppose $f \in F_i$, and $f[\alpha] = (\alpha_0, \alpha_1, \ldots, \alpha_{t-i}, \gamma_1, \ldots, \gamma_i)$. Pick any

g_b; since γ_1 is adjacent to α_{t-1}, $g_b^i(\gamma_1)$ is adjacent to $g_b^i(\alpha_{t-1}) = \alpha_t$, and so $g_b^{-i}(\gamma_1) = v^{(a)}$ for some $a \in \{1, 2, \ldots, l\}$. Then

$$g_a^{-i} g_b^i f[\alpha] = (\alpha_0, \alpha_1, \ldots, \alpha_{t-i+1}, \delta_2, \ldots, \delta_i) \quad \text{say.}$$

By applying the same method with i replaced by $i - 1$, we can find an automorphism $g_c^{-(i-1)} g_d^{i-1}$, which belongs to both Y_{i-1} and Y_i, and takes δ_2 to α_{t-i+2} while fixing $\alpha_0, \alpha_1, \ldots, \alpha_{t-i+1}$. Continuing in this way, we construct g in Y_i such that $gf[\alpha] = [\alpha]$, that is, gf is in F_0. Consequently f is in the group generated by Y_i and F_0. Conversely, both Y_i and F_0 are contained in F_i so we have the result. $\square$

All members of the sets $Y_0, Y_1, \ldots, Y_t$ fix the vertex α_0 and so belong to F_t, the stabilizer of α_0; further, we have shown that F_t is generated by Y_t and F_0. In the case of Y_{t+1}, we note that this set contains some automorphisms not fixing α_0, and we may ask whether Y_{t+1} and F_0 suffice to generate the entire automorphism group G. The following proposition shows that the answer is 'yes', unless the graph is bipartite. The reason why bipartite graphs are exceptional in this respect is that if Γ is a symmetric bipartite graph in which $V\Gamma$ is partitioned into two colour-classes V_1 and V_2, then the automorphisms which fix V_1 and V_2 setwise form a subgroup of index two in $\text{Aut}(\Gamma)$. We say that this subgroup *preserves the bipartition*.

Proposition 17.8 *Let Γ be a t-transitive graph with $t \geq 2$ and girth greater than 3. Let G^* denote the subgroup of $G = \text{Aut}(\Gamma)$ generated by Y_{t+1} and F_0. Then either (1) $G^* = G$; or (2) Γ is bipartite, $|G : G^*| = 2$, and G^* is the subgroup of G preserving the bipartition.*

Proof Let u be any vertex of Γ such that $\partial(u, \alpha_0) = 2$; we show first that there is some g^* in G^* taking α_0 to u. Since the girth of Γ is greater than 3, the vertices $v^{(a)} = g_a^{t+1}(\alpha_0)$ and $v^{(b)} = g_b^{t+1}(\alpha_0)$ satisfy $\partial(v^{(a)}, v^{(b)}) = 2$. Consequently, the distance between α_0 and $g_a^{-(t+1)} g_b^{t+1}(\alpha_0)$ is also 2. Now G^* contains F_t (since the latter is generated by Y_t, which is a subset of Y_{t+1}, and F_0), and F_t is transitive on the 2-arcs which begin at α_0 (since $t \geq 2$). Thus G^* contains an automorphism f fixing α_0 and taking $g_a^{-(t+1)} g_b^{t+1}(\alpha_0)$ to u, and $g^* = f g_a^{-(t+1)} g_b^{t+1}$ takes α_0 to u.

Let U denote the orbit of α_0 under the action of G^*. U contains all vertices whose distance from α_0 is two, and consequently all vertices whose distance from α_0 is even. If $U = V\Gamma$, then G^* is transitive on $V\Gamma$, and since it contains F_t, the stabilizer of the vertex α_0 in (G^*) is F_t. Thus $|G^*| = |V\Gamma||F_t| = |G|$, and so $G^* = G$. If $U \neq V$, then U

consists precisely of those vertices whose distance from α_0 is even, and Γ is bipartite, with colour-classes U and $V\Gamma \setminus U$. Since G^* fixes them setwise, G^* is the subgroup of G preserving the bipartition. □

We remark that the only connected graphs of girth three whose automorphism group is transitive on 2-arcs are the complete graphs. Thus the girth constraint in Proposition 17.8 is not very restrictive.

In the next chapter we shall specialize the results of Propositions 17.7 and 17.8 to 3-regular graphs; our results will lead to very precise information about the stabilizer sequence.

Additional Results

17a *The significance of the condition* $t \geq 2$ In **16d** we observed that the vertex-stabilizer G_v has a normal subgroup L_v such that G_v/L_v is a group of permutations of the vertices adjacent to v. In the case of a symmetric graph with $t \geq 2$ this group of permutations is doubly-transitive. Since all doubly-transitive permutation groups are 'known', this observation links the problem of classifying symmetric graphs with the classification theorems of group theory. See also **17f** and **17g**.

17b *The stabilizer of an edge-neighbourhood*   Suppose that Γ is a symmetric graph of degree k with $G = \text{Aut}(\Gamma)$. For any edge $\{v, w\}$ of Γ, define $G_{vw} = G_v \cap G_w$, $L_{vw} = L_v \cap L_w$, where L_v and L_w are the stabilizers of the respective vertex-neighbourhoods, as defined in **16d**. Then we have the following subgroup relationships among these groups.
(a) L_v is a normal subgroup of G_v and G_{vw};
(b) L_{vw} is a normal subgroup of L_v and G_{vw}.
It follows from standard theorems of group theory that

$$\frac{L_v}{L_{vw}} \approx \frac{L_v L_w}{L_w}$$

and $L_v L_w / L_w$ is a normal subgroup of G_{vw}/L_w. The last group is a group of permutations of the neighbours of w, fixing v. Thus we have $|L_v : L_{vw}| \leq (k-1)!$, and

$$|G_v| \leq k!(k-1)!|L_{vw}|.$$

17c *The full automorphism group of* $K_{n,n}$ It is clear that the graph $K_{n,n}$ has at least $2(n!)^2$ automorphisms. Simple arguments suffice to show that there are no others, but, for the sake of example, we can use **17b**. In this case the neighbourhood of an edge is the whole graph, so $L_{vw} = 1$. It follows that

$$|G| \leq 2n|G_v| \leq 2n\, n!\, (n-1)! = 2\, (n!)^2.$$

17d *The automorphism group of O_k* A more substantial application of **17b** shows that the symmetric group S_{2k-1} is the full automorphism group of O_k. When $k \geq 3$ every 3-arc in O_k determines a unique 6-cycle, and it follows from this that if $g \in L_{vw}$ then $g \in L_{wx}$ for all vertices x adjacent to w. Hence $L_{vw} = 1$, and the order of the full automorphism group is at most

$$k! \, (k-1)! \binom{2k-1}{k-1} = (2k-1)!.$$

An alternative proof using the Erdős–Ko–Rado theorem may be found in Biggs (1979).

17e *The stabilizer sequence for odd graphs* The odd graphs O_k are 3-transitive, for all $k \geq 3$. The stabilizer sequence is

$$G = S_{2k-1}, \quad F_3 = S_k \times S_{k-1}, \quad F_2 = S_{k-1} \times S_{k-1},$$

$$F_1 = S_{k-1} \times S_{k-2}, \quad F_0 = S_{k-2} \times S_{k-2}.$$

17f *L_{vw} is a p-group* (Gardiner 1973) For any t-transitive graph with $t \geq 2$ the edge-neighbourhood stabilizer L_{vw} is a p-group, for some prime p. If $t \geq 4$ and the degree is $p+1$ it follows that the order of a vertex-stabilizer G_v is $(p+1)p^{t-1}m$, where $t = 4, 5$ or 7 and m is a divisor of $(p-1)^2$.

17g *There are no 8-transitive graphs* Weiss (1983) extended the results of Gardiner and others, and using the classification theorems of group theory he showed that there are no finite graphs (apart from the cycles) for which a group of automorphisms can act transitively on the t-arcs for $t \geq 8$. 7-transitive graphs do exist: the smallest is a 4-regular graph with 728 vertices [BCN, p. 222].

17h *Symmetric cycles* A cycle with vertices $v_0, v_1, \ldots, v_{l-1}$ in a graph Γ is *symmetric* if there is an automorphism g of Γ such that $g(v_i) = v_{i+1}$, where the subscripts are taken modulo l. J.H. Conway observed that in a symmetric graph of degree k the symmetric cycles fall into $k-1$ equivalence classes under the action of the automorphism group. The details may be found in Biggs (1981a). For example, the two classes in Petersen's graph contain 5-cycles and 6-cycles, and in general, the classes in O_k have lengths $6, 10, \ldots, 4k-6$ and $2k-1$.

18

Symmetric graphs of degree three

In this chapter we shall use the traditional term *cubic graph* to denote a simple, connected graph which is regular of degree three. As we shall see, the theory of symmetric cubic graphs is full of strange delights.

Suppose that Γ is a t-transitive graph so that, by definition, $\mathrm{Aut}(\Gamma)$ is transitive on the t-arcs of Γ but not transitive on the $(t+1)$-arcs of Γ. The distinctive feature of the cubic case is that $\mathrm{Aut}(\Gamma)$ acts *regularly* on the t-arcs.

Proposition 18.1   *Let $[\alpha]$ be a t-arc in a cubic t-transitive graph Γ. Then an automorphism of Γ which fixes $[\alpha]$ must be the identity.*

Proof   Suppose f is an automorphism fixing each vertex $\alpha_0, \alpha_1, \ldots, \alpha_t$. If f is not the identity, then f does not fix all t-arcs in Γ. It follows from Lemma 17.4 that there is some t-arc $[\beta]$ such that f fixes $[\beta]$, but f does not fix both successors of $[\beta]$. Clearly, if $\beta_{t-1}, u^{(1)}, u^{(2)}$ are the vertices adjacent to β_t, then f must interchange $u^{(1)}$ and $u^{(2)}$. Let $w \neq \beta_1$ be a vertex adjacent to β_0. Since Γ is t-transitive there is an automorphism $h \in \mathrm{Aut}(\Gamma)$ taking the t-arc $(w, \beta_0, \ldots, \beta_{t-1})$ to $[\beta]$, and we may suppose the notation chosen so that $h(\beta_t) = u^{(1)}$. Then h and fh are automorphisms of Γ taking the $(t+1)$-arc $[w.\beta]$ to its two successors, and, by Theorem 17.5, $\mathrm{Aut}(\Gamma)$ is transitive on $(t+1)$-arcs. This contradicts our hypothesis, and so we must have $f = 1$. $\square$

From now on we shall suppose that we are dealing with a cubic t-transitive graph Γ, and that we have chosen an arbitrary t-arc $[\alpha]$ in Γ.

If the stabilizer sequence of this t-arc is
$$\text{Aut}(\Gamma) = G > F_t > F_{t-1} > \ldots > F_0,$$
then Proposition 18.1 implies that $|F_0| = 1$. Consequently we know the orders of all the groups in the stabilizer sequence:
$$|F_i| = 2^i \qquad (0 \le i \le t-1),$$
$$|F_t| = 3 \times 2^{t-1},$$
$$|G| = n \times 3 \times 2^{t-1} \quad (n = |V\Gamma|).$$

The structure of these groups can be elucidated by investigating certain sets of generators for them. These generators are derived from the sets Y_i defined for the general case in Chapter 17. Let $\alpha_{t-1}, v^{(1)}, v^{(2)}$ be the vertices adjacent to α_t, and let g_r $(r = 1, 2)$ denote automorphisms taking $[\alpha]$ to $(\alpha_1, \alpha_2, \ldots, \alpha_t, v^{(r)})$. We shall use the following notation:
$$g = g_1, \quad x_0 = g_1^{-1} g_2, \quad x_i = g^{-i} x_0 g^i \quad (i = 1, 2, \ldots, t).$$
The effect of these automorphisms on the basic t-arc $[\alpha]$ is indicated in Figure 13. We note that these automorphisms are unique, as a consequence of Proposition 18.1.

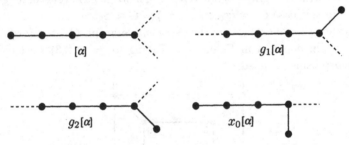

Figure 13: the effect of g_1, g_2 and x_0 on $[\alpha]$.

In this chapter, $\langle X \rangle$ will denote the subgroup of $\text{Aut}(\Gamma)$ generated by the set X.

Proposition 18.2 *The stabilizer sequence of a cubic t-transitive graph with $t \ge 2$ has the following properties:*
(1) $F_i = \langle x_0, x_1, \ldots, x_{i-1} \rangle$ *for* $i = 1, 2, \ldots, t$;
(2) *if* $G^* = \langle x_0, x_1, \ldots, x_t \rangle$ *then* $|G : G^*| \le 2$;
(3) $G = \langle x_0, g \rangle$.

Proof We shall use the notation and results of Propositions 17.7 and 17.8. In the cubic case we have $F_0 = 1$, and the set Y_i consists of the elements $g_1^{-j} g_2^j$ and their inverses $g_2^{-j} g_1^j$ for $1 \le j \le i$.
(1) It follows from part (2) of Proposition 17.7 that $F_i = \langle Y_i \rangle$. Now
$$g_1^{-j} g_2^j = x_{j-1} g_1^{-(j-1)} g_2^{j-1} = x_{j-1} x_{j-2} \ldots x_0,$$

and so $F_i = \langle x_0, x_1, \ldots, x_{i-1} \rangle$.

(2) It follows from Proposition 17.8 that the group $G^* = \langle Y_{t+1} \rangle$, that is $\langle x_0, x_1, \ldots, x_t \rangle$, is a subgroup of index 1 or 2 in G, provided that the girth of Γ is greater than three. If the girth is three, then it is easy to see that the only possibility is $t = 2, \Gamma = K_4$, and we may verify the conclusion explicitly in that case.

(3) If $G = G^*$, then $\langle x_0, g \rangle$ contains $\langle x_0, x_1, \ldots, x_t \rangle = G^* = G$. If $|G : G^*| = 2$, then Γ is bipartite, and each element g^* of G^* moves vertices of Γ through an even distance in Γ. But the element $g = g_1$ moves some vertices to adjacent vertices, and so $g \notin G^*$. Thus, adjoining g to G^* must enlarge the group, and since G^* is a maximal subgroup of G (because it has index 2) we have $\langle G^*, g \rangle = \langle x_0, g \rangle = G$. $\square$

In the previous chapter we considered Petersen's graph, obtaining for the 3-arc $[\alpha] = (12, 34, 15, 23)$ the automorphisms $g_1 = (13)(245), g_2 = (13524)$. Hence

$$x_0 = (34), \quad x_1 = (12), \quad x_2 = (35), \quad x_3 = (14).$$

We know that this graph is not bipartite, since it has cycles of length 5, and so in this case $G^* = \langle x_0, x_1, x_2, x_3 \rangle = G \approx S_5$.

Another simple example is the 2-transitive graph Q_3, the (ordinary) cube graph, depicted in Figure 14. Taking $[\alpha] = (1,2,3)$ we have the automorphisms as listed.

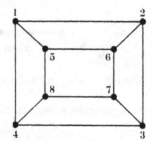

Figure 14: the cube graph Q_3.

$$g_1 = (1234)(5678), \quad g_2 = (123785)(46),$$

$$x_0 = (36)(45), \quad x_1 = (16)(47), \quad x_2 = (18)(27).$$

In this case the graph is bipartite and $G^* = \langle x_0, x_1, x_2 \rangle$ preserves the bipartition

$$VQ_3 = \{1, 3, 5, 7\} \cup \{2, 4, 6, 8\}.$$

It follows that $|G : G^*| = 2$.

The main result on t-transitive cubic graphs is that there are no finite examples with $t > 5$. The proof of this very important result is due to Tutte (1947a), with later improvements by Sims (1967) and Djoković (1972). Following these authors we shall obtain the result as an algebraic consequence of the presentation of the stabilizer sequence given in Proposition 18.2. A rather more streamlined proof, using 'geometrical' arguments to replace some of the algebraic calculations, has been given by Weiss (1974).

We shall suppose that $t \geq 4$, as this assumption helps to avoid vacuous statements. We observe that each generator x_i $(i \geq 0)$ is an involution, and that each element of F_i $(1 \leq i \leq t - 1)$ has a unique expression in the form

$$x_\rho x_\sigma \ldots x_\tau, \quad \text{where} \quad 0 \leq \rho < \sigma < \ldots < \tau \leq i - 1,$$

where we allow the empty set of subscripts to represent the identity element. The uniqueness of the expression is a consequence of the fact that there are 2^i such expressions, and $|F_i| = 2^i$ for $1 \leq i \leq t - 1$.

The key idea is to determine which stabilizers are abelian and which are non-abelian. It is immediate that F_1 and F_2 are abelian, since $|F_1| = 2$ and $|F_2| = 4$. Let λ denote the largest natural number such that F_λ is abelian.

Proposition 18.3 *If* $t \geq 4$, *then* $2 \leq \lambda < \frac{1}{2}(t + 2)$.

Proof We have already remarked that $\lambda \geq 2$. Suppose that $F_\lambda = \langle x_0, \ldots, x_{\lambda-1} \rangle$ is abelian, so that its conjugate $g^{-t+\lambda-1} F_\lambda g^{t-\lambda+1}$, that is $\langle x_{t-\lambda+1}, \ldots, x_t \rangle$, is also abelian. If

$$\lambda - 1 \geq t - \lambda + 1,$$

then both these groups contain $x_{\lambda-1}$, and together they generate G^*; hence $x_{\lambda-1}$ commutes with every element of G^*. Now $g^2 \in G^*$ (since $g \in G$ and $|G : G^*| \leq 2$) and so

$$x_{\lambda-1} = g^{-2} x_{\lambda-1} g^2 = x_{\lambda+1},$$

whence $x_0 = x_2$. This is false, given $t \geq 4$, since $|F_3| > |F_2|$, and so we must have

$$\lambda - 1 < t - \lambda + 1, \quad \text{that is} \quad \lambda < \frac{1}{2}(t + 2),$$

as claimed. $\qquad\qquad\qquad\qquad\qquad\qquad\qquad\qquad\qquad\qquad\quad$ $\square$

Proposition 18.3 gives an upper bound for λ in terms of t. We shall find a lower bound of the same kind by means of arguments involving the commutators $[a, b] = a^{-1} b^{-1} a b$ of the canonical generators x_i. Note that since these generators are involutions, we have

$$[x_i, x_j] = (x_i x_j)^2.$$

Lemma 18.4   *The generators x_i satisfy the following conditions.*
(1) $[x_i, x_j] = 1$ *if* $|j - i| < \lambda$, *but* $[x_i, x_j] \neq 1$ *if* $|j - i| = \lambda$.
(2) *The centre of* $F_j = \langle x_0, \ldots, x_{j-1} \rangle$ *is the group* $\langle x_{j-\lambda}, \ldots, x_{\lambda-1} \rangle$ ($\lambda \leq j < 2\lambda$).
(3) *The commutator subgroup of* F_{i+1} *is a subgroup of* $\langle x_1, \ldots, x_{i-1} \rangle$ $= g^{-1}F_{i-1}g$ ($1 \leq i \leq t - 2$).

Proof   (1) We may suppose without loss that $j > i$; then $[x_i, x_j] = g^{-i}[x_0, x_{j-i}]g^i$ and so $[x_i, x_j] = 1$ if and only if x_0 and x_{j-i} commute. The result follows from the fact that $F_\lambda = \langle x_0, \ldots, x_{\lambda-1} \rangle$ is the largest abelian stabilizer.

(2) If the non-identity element x of F_j is written in the form

$$x_\rho x_\sigma \ldots x_\tau \quad (0 \leq \rho < \sigma < \ldots < \tau \leq j - 1),$$

then x does not commute with $x_{\rho+\lambda}$. Further, if $\rho + \lambda < j$ then $x_{\rho+\lambda}$ belongs to F_j. Similarly, x does not commute with $x_{\tau-\lambda}$, and if $\tau - \lambda > -1$, then $x_{\tau-\lambda}$ belongs to F_j. Thus, if x is in the centre of F_j then $\rho \geq j - \lambda$ and $\tau \leq \lambda - 1$, so that x is in $\langle x_{j-\lambda}, \ldots, x_{\lambda-1} \rangle$. Conversely, it follows from (1) that every element of this group is in the centre of F_j.

(3) Provided that $1 \leq i \leq t - 2$, the groups $F_i = \langle x_0, \ldots, x_{i-1} \rangle$ and $g^{-1}F_i g = \langle x_1, \ldots, x_i \rangle$ are different, and they are both of index two in F_{i+1}, and consequently normal in F_{i+1}. Thus their intersection $\langle x_1, \ldots, x_{i-1} \rangle = g^{-1}F_{i-1}g$ is normal in F_{i+1}, and the quotient group $F_{i+1}/(g^{-1}F_{i-1}g)$ is abelian, since it has order 4. Hence the commutator subgroup of F_{i+1} is contained in $g^{-1}F_{i-1}g$.   □

Since $[x_0, x_\lambda]$ belongs to the commutator subgroup of $F_{\lambda-1}$, it follows (from part (3) of Lemma 18.4 with $i = \lambda$) that $[x_0, x_\lambda]$ belongs to the group $\langle x_1, \ldots, x_{\lambda-1} \rangle$. In other words, there is a unique expression

$$[x_0, x_\lambda] = x_\mu \ldots x_\nu \quad (1 \leq \mu \leq \nu \leq \lambda - 1).$$

Lemma 18.5   *With the above notation, we have:*

$$(1) \quad \mu + \lambda \geq t - 1; \quad (2) \quad 2\lambda - \nu \geq t - 1.$$

Proof   (1) Suppose that $\mu + \lambda \leq t - 2$. Then (by part (3) of Lemma 18.4) the element $[x_0, x_{\mu+\lambda}]$ of the commutator subgroup of $F_{\mu+\lambda+1}$ is contained in $\langle x_1, \ldots, x_{\mu+\lambda-1} \rangle$. The centre of $\langle x_1, \ldots, x_{\mu+\lambda-1} \rangle$ is the group $\langle x_\mu, \ldots, x_\lambda \rangle$, and since this contains both x_λ and $[x_0, x_\lambda]$ it follows that $[x_0, x_{\mu+\lambda}]$ commutes with x_λ and with $[x_0, x_\lambda]$. Also x_λ commutes with $x_{\mu+\lambda}$, since $\mu \leq \lambda - 1$. Hence we have the following calculation:

$$x_{\mu+\lambda}^{-1}[x_0, x_\lambda]x_{\mu+\lambda} = [x_{\mu+\lambda}^{-1}x_0x_{\mu+\lambda}, x_\lambda]$$
$$= [x_0[x_0, x_{\mu+\lambda}], x_\lambda]$$
$$= [x_0, x_{\mu+\lambda}]^{-1}[x_0, x_\lambda][x_0, x_{\mu+\lambda}][[x_0, x_{\mu+\lambda}], x_\lambda]$$
$$= [x_0, x_\lambda].$$

This implies that $x_{\mu+\lambda}$ commutes with $[x_0, x_\lambda] = x_\mu \ldots x_\nu$. But this is false, since $x_{\mu+\lambda}$ does not commute with x_μ but does commute with any other term in the expression for $[x_0, x_\lambda]$. Thus our hypothesis was wrong, and $\mu + \lambda \geq t - 1$.

(2) If $2\lambda - \nu \leq t - 2$, then using arguments parallel to those in (1), we may prove that $[x_{2\lambda-\nu}, x_0]$ commutes with $x_{\lambda-\nu}$ and with $[x_{\lambda-\nu}, x_{2\lambda-\nu}]$; also $x_{\lambda-\nu}$ commutes with x_0, since $\nu \geq 1$. A calculation like that in (1) then implies that x_0 commutes with

$$[x_{\lambda-\nu}, x_{2\lambda-\nu}] = x_{\mu+\lambda-\nu} \ldots x_\lambda,$$

which is false. Hence $2\lambda - \nu \geq t - 1$.   □

Theorem 18.6 (Tutte 1947a)   *There is no finite t-transitive cubic graph with $t > 5$.*

Proof   If t is at least four, then Proposition 18.3 tells us that $\lambda < \frac{1}{2}(t+2)$. However, the results of Lemma 18.5 show that $t - 1 - \lambda \leq \mu \leq \nu \leq 2\lambda - t + 1$; that is, $\lambda \geq \frac{2}{3}(t-1)$. Now, if $t \geq 4$ there is an integer λ such that

$$\frac{2}{3}(t-1) \leq \lambda < \frac{1}{2}(t+2)$$

only when $t = 4, 5, 7$. It remains to exclude the possibility $t = 7$, which is done by means of the following special argument.

If Γ is a 7-transitive cubic graph, then the inequalities for λ, μ and ν imply that $\lambda = 4$, $\mu = \nu = 2$; thus $[x_0, x_4] = x_2$. Also, by part (3) of Lemma 18.4, $[x_0, x_5]$ belongs to the group $\langle x_1, x_2, x_3, x_4 \rangle$. If the standard expression for $[x_0, x_5]$ actually contains x_4, then we can write $[x_0, x_5] = hx_4$, where $h \in \langle x_1, x_2, x_3 \rangle$ so that h commutes with x_0 and x_4. Hence

$$x_2 = [x_0, x_4] = (x_0x_4)^2 = (x_0hx_4)^2 = (x_0(x_0x_5)^2)^2$$
$$= (x_5x_0x_5)^2 = x_5x_0^2x_5 = 1.$$

Since this is absurd, $[x_0, x_5] = (x_0x_5)^2$ must belong to $\langle x_1, x_2, x_3 \rangle$.

Now the original definitions show that x_1, x_2, and x_3 fix the vertex α_3 of the 7-arc $[\alpha]$, and so $x_0x_5(\alpha_3) = x_5x_0(\alpha_3) = x_5(\alpha_3)$. That is, x_0 fixes $x_5(\alpha_3)$. Further, since x_5 fixes α_1 but not α_2 we have a 7-arc $[\theta] = (x_5(\alpha_3), x_5(\alpha_2), \alpha_1, \alpha_2, \alpha_3, \alpha_4, \alpha_5, \alpha_6)$ in Γ. The three vertices adjacent to α_1 are α_0, α_2 and $x_5(\alpha_2)$, and since x_0 fixes α_0, α_1 and α_2 it must fix $x_5(\alpha_2)$ also. Consequently x_0 fixes the whole 7-arc $[\theta]$, and this contradicts Proposition 18.1. Hence $t = 7$ cannot occur. $\square$

Goldschmidt (1980) proved an important extension of this result.

The simplest example of a 5-transitive cubic graph is constructed as follows. Let the symmetric group S_6 act on the 6 symbols $\{a, b, c, d, e, f\}$, and take the vertices of a graph Ω to be the 15 permutations of shape (ab) and the 15 permutations of shape $(ab)(cd)(ef)$. Join two vertices by an edge if and only if the corresponding permutations have different shape and they commute. For instance, (ab) is joined to the vertices $(ab)(cd)(ef)$, $(ab)(ce)(df)$ and $(ab)(cf)(de)$, while $(ab)(cd)(ef)$ is joined to $(ab), (cd)$ and (ef). Clearly, any automorphism of the group S_6 is an automorphism of Ω, and so

$$|\mathrm{Aut}(\Omega)| = |\mathrm{Aut} S_6| = 1440 = 30 \times 3 \times 2^4,$$

as we expect for a 5-transitive cubic graph with 30 vertices. We can verify that Ω is indeed 5-transitive by working out generators in terms of the following 5-arc:

$$(ab), \quad (ab)(cd)(ef), \quad (cd), \quad (ae)(bf)(cd), \quad (ae), \quad (ae)(bd)(cf).$$

If π is an element of S_6, denote the corresponding inner automorphism (conjugation) of S_6 by $|\pi|$. Then the generators for the stabilizer sequence may be chosen as follows:

$$x_0 = |(cd)|, \quad x_1 = |(ab)(cd)(ef)|, \quad x_2 = |(ab)|,$$
$$x_3 = |(ab)(cf)(de)|, \quad x_4 = |(cf)|.$$

The groups which occur in the stabilizer sequence are

$$F_5 = S_4 \times \mathbb{Z}_2, \quad F_4 = D_8 \times \mathbb{Z}_2, \quad F_3 = (\mathbb{Z}_2)^3,$$
$$F_2 = (\mathbb{Z}_2)^2, \quad F_1 = \mathbb{Z}_2.$$

Finally, we may choose x_5 so that $G^* = \langle x_0, \ldots, x_5 \rangle$ is isomorphic to S_6, and so $|G : G^*| = 2$ in accordance with the fact that the graph is bipartite.

Additional Results

18a *A non-bipartite 5-transitive cubic graph*   A 5-transitive cubic graph with 234 vertices, which is not bipartite, can be constructed as follows. The vertices correspond to the 234 triangles in $PG(2,3)$ and two vertices are adjacent whenever the corresponding triangles have one common point and their remaining four points are distinct and collinear. The automorphism group is the group Aut $PSL(3,3)$, of order $11232 = 234 \times 3 \times 2^4$.

18b *The sextet graphs* (Biggs and Hoare 1983)   Let q be an odd prime power. Define a *duet* to be an unordered pair of points ab on the projective line $PG(1,q) = GF(q) \cup \{\infty\}$, and a *quartet* to be an unordered pair of duets $\{ab \mid cd\}$ such that the cross-ratio

$$\frac{(a-c)(b-d)}{(a-d)(b-c)} = -1.$$

(The usual conventions about ∞ apply here.) A *sextet* is an unordered triple of duets $\{ab \mid cd \mid ef\}$ such that each of $\{ab \mid cd\}$, $\{cd \mid ef\}$ and $\{ef \mid ab\}$ is a quartet. There are $q(q^2-1)/24$ sextets if $q \equiv 1 \pmod 4$, and none if $q \equiv 3 \pmod 4$.

When $q \equiv 1 \pmod 8$ it is possible to define 'adjacency' of sextets in such a way that each sextet is adjacent to three others. Thus we obtain a regular graph $\Sigma(q)$ of degree 3, whose components $\Sigma_0(q)$ are all isomorphic. The *sextet graph* $S(p)$ is defined to be $\Sigma_0(p)$ if $p \equiv 1 \pmod 8$ and $\Sigma_0(p^2)$ if $p \equiv 3, 5, 7 \pmod 8$.

The sextet graphs $S(p)$ so defined form an infinite family of cubic graphs, one for each odd prime p. The graph $S(p)$ is 5-transitive when $p \equiv 3$ or $5 \pmod 8$, and 4-transitive otherwise. The order of $S(p)$ depends on the congruence class of p modulo 16, as follows:

$$n = \frac{1}{48}p(p^2-1) \quad \text{when} \quad p \equiv 1, 15 \pmod{16};$$

$$n = \frac{1}{24}p(p^2-1) \quad \text{when} \quad p \equiv 7, 9 \pmod{16};$$

$$n = \frac{1}{24}p^2(p^4-1) \quad \text{when} \quad p \equiv 3, 5, 11, 13 \pmod{16}.$$

The group $\text{Aut}S(p)$ is $PSL(2,p)$, $PGL(2,p)$, $P\Gamma L(2,p^2)$ in the respective cases. The two smallest 5-transitive sextet graphs are $S(3)$, which is isomorphic to the graph Ω described above, and $S(5)$, which is a graph with 650 vertices.

18c *Conway's presentations and the seven types*   Given an arbitrary t-arc $[\alpha]$, let a and b denote the automorphisms taking $[\alpha]$ to its successors (so $a = g_1$ and $b = g_2$ in the notation described at the beginning of this chapter). Also, let σ be the automorphism which reverses $[\alpha]$; that is,

$$\sigma(\alpha_i) = \alpha_{t-i} \quad (0 \le i \le t).$$

Since we know that $\text{Aut}(\Gamma)$ acts regularly on the t-arcs, it follows that σ^2 is the identity and $\sigma a \sigma$ is either a^{-1} or b^{-1}. We denote the case when $\sigma a \sigma = a^{-1}$ by t^+ and the case when $\sigma a \sigma = b^{-1}$ by t^-. It turns out that the t^+ case can occur only when $t = 2, 3, 4, 5$ and the t^- case only when $t = 1, 2, 4$.

In each of the cases it can be shown, by analysis of the action of suitable combinations of a, b and σ on $[\alpha]$, that certain relations must hold in Aut(Γ). For example, in the 2^+ case these relations are:

$$\sigma^2 = 1, \quad (\sigma a)^2 = 1, \quad (\sigma b)^2 = 1, \quad (a^{-1}b)^2 = 1, \quad ab\sigma a^2 = b^2.$$

In the 5^+ case they are:

$$\sigma^2 = 1, \quad (\sigma a)^2 = 1, \quad (\sigma b)^2 = 1, \quad (a^{-1}b)^2 = 1, \quad (a^{-2}b^2)^2 = 1,$$

$$(a^{-3}b^3)^2 = 1, \quad a^4 b^{-4} a^4 = ba^2 b, \quad a^4 b\sigma a^5 = ba^3 b.$$

Let us denote the groups generated by a, b and σ, subject to the appropriate relations, by

$$G_2^+, G_3^+, G_4^+, G_5^+, G_1^-, G_2^-, G_4^-.$$

Each of these groups is an infinite group of automorphisms of the infinite cubic tree T_3, acting regularly on the t-arcs, for the relevant value of t, and they are the only such groups, up to conjugacy in Aut(T_3). More detailed information about the seven groups, using different presentations, is given by Djoković and Miller (1980), and Conder and Lorimer (1989).

18d *Finite cubic graphs and groups*　　Any group acting regularly on the t-arcs of a finite cubic graph Γ is a quotient of one of the seven groups in **18c**. The quotient is defined by adding relations which represent cycles in Γ, a cycle of length l in Γ corresponding to a word of length l in a and b which represents the identity. For example, adding the relation $a^4 = 1$ to the relations for G_2^+ defines a group $G_2^+(a^4)$. This is the group of the cube Q_3, as can be verified by showing that the permutations

$$a = (1234)(5678), \quad b = (123785)(46), \quad \sigma = (13)(57),$$

satisfy the defining relations for $G_2^+(a^4)$ and represent automorphisms of Q_3 acting in the prescribed way on the 2-arc $(1, 2, 3)$ (see Figure 14).

18e *Coset enumeration*　　In the notation of the Conway presentations, the stabilizer of a t-arc is

$$F_t = \langle a^{-i}b^i \mid i = 1, 2, \ldots, t \rangle.$$

If G is a quotient of G_t^+ or G_t^-, then the index $|G : F_t|$ is the cardinality of a cubic graph for which G is a t-transitive group of automorphisms. The index may be finite or infinite, but if it is finite the method of coset enumeration will (in principle) determine its value. This is a powerful method for constructing finite t-transitive cubic graphs. See Biggs (1984a) for further details.

18f *The structure of a stabilizer sequence*　　The groups occurring in the stabilizer sequence are determined up to isomorphism, as in the following table. (Note that when $t = 2, 4$ both the t^+ and t^- cases can occur, but the abstract groups are the same.)

t	F_1	F_2	F_3	F_4	F_5
1	$\mathbb{Z}_3$				
2	$\mathbb{Z}_2$	S_3			
3	$\mathbb{Z}_2$	$(\mathbb{Z}_2)^2$	D_{12}		
4	$\mathbb{Z}_2$	$(\mathbb{Z}_2)^2$	D_8	S_4	
5	$\mathbb{Z}_2$	$(\mathbb{Z}_2)^2$	$(\mathbb{Z}_2)^3$	$D_8 \times \mathbb{Z}_2$	$S_4 \times \mathbb{Z}_2$.

18g *Symmetric* **Y** *and* **H** *graphs*   Let **Y** and **H** denote the trees whose pictorial representations correspond to the respective letters. Both of these trees have vertices of degree 1 (leaves) and 3 only. Given any such tree T we can form an *expansion* of T by taking a number n of disjoint copies of T and joining each set of corresponding leaves by a cycle of length n; each cycle has a constant 'step', and different cycles will in general have different steps. For example, when $T = K_2$, we get the graphs $P(n, t)$ described in **15e** by joining one set of leaves with step 1 and the other set with step t.

Clearly an expansion of T is a cubic graph. The result quoted in **15e** implies that only seven expansions of K_2 are symmetric. Horton and Bouwer (1991) showed that there are only six other expansions which are symmetric. Four of them are expansions of **Y**: $n = 7$, steps 1,2,4; $n = 14$, steps 1,3,5; $n = 28$, steps 1,3,9; $n = 56$, steps 1,9,25. The other two are expansions of **H**: $n = 17$, steps 1,2,4,8; and $n = 34$, steps 1,9,13,15.

18h *Foster's census of symmetric cubic graphs*   (More details and bibliographical references relating to the following sketch are given by Bouwer (1988).) In 1920 two electrical engineers, G.A. Campbell and R.M. Foster, wrote a paper in which the graph $K_{3,3}$ was used in the context of 'telephone substation and repeater circuits'. Twelve years later Foster published drawings of nine symmetric cubic graphs. He continued to work on the subject, and in 1966 he spoke at a conference at the University of Waterloo, where he distributed a mimeographed list of such graphs with up to 400 vertices. In 1988, when Foster was just 92, Bouwer and his colleagues published Foster's census for graphs with up to 512 vertices. Remarkably, only five graphs (out of 198) are known to have been missed by Foster, and workers in this field are convinced that there can be very few others, if any.

The graphs with $n \le 30$ vertices are as follows.

K_4, $K_{3,3}$, Q_3, Petersen's graph, Heawood's graph $S(7)$, $P(8,3)$ (see **15e**), the Pappus graph (see **19h**), $P(10,3)$ or the Desargues graph (see **19b**), the dodecahedron, $P(12,5)$, $\mathbf{Y}(7:1,2,4)$ (see **18g**), and $\Omega = S(3)$.

18i *All 5-transitive cubic graphs with less than 5000 vertices*   Coset enumerations based on the Conway presentations and other techniques have established that the following list of 5-transitive cubic graphs with $n \leq 5000$ vertices is 'almost certainly' complete.

$n = 30$: the sextet graph $S(3)$, group $G_5^+(a^8)$;

$n = 90$: a threefold cover of $S(3)$ (see **19c**), group $G_5^+(b^{10})$;

$n = 234$: the graph described in **18a**, group $G_5^+(a^{13})$;

$n = 468$: a double covering of the previous graph, group $G_5^+(b^{12})$;

$n = 650$: the sextet graph $S(5)$, group $G_5^+(a^{12})$;

$n = 2352$: a graph to be described in **19e**, group $G_5^+(a^{14})$;

$n = 4704$: a double covering of the previous graph, group $G_5^+((ab)^8)$.

18j *The symmetric group S_{10} is a quotient of G_5^+* (Conder 1987)   The following permutations of $\{1,\ldots,9,X\}$ satisfy the Conway relations for G_5^+, as given in **18c**:

$$a = (12)(34675)(89X), \quad b = (1246853)(79X), \quad \sigma = (12)(34)(56)(9X).$$

Since these permutations generate the symmetric group S_{10}, it follows that there is a 5-transitive cubic graph with $10!/48 = 75600$ vertices. The graph can be constructed in a way which shows that it is closely related to the simplest 5-transitive cubic graph, the graph $\Omega = S(3)$ (Lorimer 1989).

19

The covering graph construction

In this chapter we shall study a 'covering graph' technique which, in certain circumstances, enables us to manufacture new symmetric graphs from a given one. The method was first used in this context by J.H. Conway, who used the simple version discussed in Theorem 19.5 to show that there are infinitely many connected cubic graphs which are 5-transitive. The general version given here was developed in the original 1974 edition of this book, and has since found several other applications, some of which are described in the Additional Results at the end of the chapter. The related technique of 'voltage-graphs' (see Gross 1974) is much used in the theory of graph embeddings.

We shall use the symbol $S\Gamma$ to denote the set of 1-arcs or *sides* of a graph Γ; each edge $\{u, v\}$ of Γ gives rise to two sides, (u, v) and (v, u). For any group K, we define a *K-chain* on Γ to be a function $\phi : S\Gamma \to K$ such that $\phi(u, v) = (\phi(v, u))^{-1}$ for all sides (u, v) of Γ .

Definition 19.1 The *covering graph* $\widetilde{\Gamma} = \widetilde{\Gamma}(K, \phi)$ of Γ, with respect to a given K-chain ϕ on Γ, is defined as follows. The vertex-set of $\widetilde{\Gamma}$ is $K \times V\Gamma$, and two vertices $(\kappa_1, v_1), (\kappa_2, v_2)$ are joined by an edge if and only if

$$(v_1, v_2) \in S\Gamma \quad \text{and} \quad \kappa_2 = \kappa_1 \phi(v_1, v_2).$$

It is easy to check that the definition of adjacency depends only on the unordered pair of vertices.

As an example let $\Gamma = K_4$, and let K be the group $\mathbb{Z}_2$ whose elements

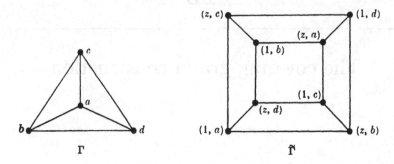

Figure 15: Q_3 as a double covering of K_4.

are 1 and z; the function ϕ which assigns z to each side of K_4 is a $\mathbb{Z}_2$-chain on K_4. The covering graph $\widetilde{\Gamma}(\mathbb{Z}_2, \phi)$ is isomorphic to the cube Q_3, as depicted in Figure 15.

Suppose that a group G acts as a group of automorphisms of a group K; that is, for each g in G we have an automorphism $\hat{g}$ of K such that the function $g \mapsto \hat{g}$ is a group homomorphism from G to $\mathrm{Aut}K$. In this situation we define the *semi-direct product* of K by G, denoted by $K\widehat{\times}G$, to be the group whose elements are the ordered pairs (κ, g), with the group operation given by

$$(\kappa_1, g_1)(\kappa_2, g_2) = (\kappa_1 \hat{g}_1(\kappa_2), g_1 g_2).$$

Let Γ be a graph, ϕ a K-chain on Γ, and let $G = \mathrm{Aut}(\Gamma)$. Then G acts on the sides of Γ by the rule $g(u, v) = (g(u), g(v))$, and we may postulate a special relationship between the action of G on K and its action on $S\Gamma$.

Definition 19.2 The K-chain ϕ is *compatible* with the given actions of G on K and $S\Gamma$, if the following diagram is commutative for each g in G.

$$
\begin{array}{ccc}
S\Gamma & \xrightarrow{\phi} & K \\
\downarrow{g} & & \downarrow{\hat{g}} \\
S\Gamma & \xrightarrow{\phi} & K
\end{array}
$$

Proposition 19.3 *Suppose that Γ is a graph whose automorphism group $G = \mathrm{Aut}(\Gamma)$ acts as a group of automorphisms of a group K. Suppose further that there is a K-chain ϕ on Γ which is compatible with the actions of G on K and $S\Gamma$. Then the semi-direct product $K\widehat{\times}G$ is a group of automorphisms of the covering graph $\widetilde{\Gamma} = \widetilde{\Gamma}(K, \phi)$.*

Proof Define the effect of an element (κ, g) of $K \hat{\times} G$ on a vertex (κ', v) of $\tilde{\Gamma}$ by the rule

$$(\kappa, g)(\kappa', v) = (\kappa \hat{g}(\kappa'), g(v)).$$

Using the definition of compatibility, a simple calculation shows that this permutation of $V\tilde{\Gamma}$ is an automorphism of $\tilde{\Gamma}$. □

The usefulness of the covering graph construction lies in the fact that a much stronger version of Proposition 19.3 is true.

Proposition 19.4 *With the notation and hypotheses of Proposition 19.3, suppose also that G is transitive on the t-arcs of Γ. Then $K \hat{\times} G$ is transitive on the t-arcs of $\tilde{\Gamma}$.*

Proof Let $((\kappa_0, v_0), \ldots, (\kappa_t, v_t))$ and $((\kappa_0', v_0'), \ldots, (\kappa_t', v_t'))$ be two t-arcs in $\tilde{\Gamma}$. Then $(v_0, \ldots, v_t)$ and $(v_0', \ldots, v_t')$ are t-arcs in Γ, and so there is some g in G such that $g(v_i) = v_i'$ $(0 \le i \le t)$. Suppose we choose κ^* in K such that (κ^*, g) takes (κ_0, v_0) to (κ_0', v_0'); that is, we choose $\kappa^* = \kappa_0'(\hat{g}(\kappa_0))^{-1}$. Then we claim that (κ^*, g) takes (κ_i, v_i) to (κ_i', v_i') for all $i \in \{0, 1, \ldots, t\}$.

The claim is true when $i = 0$, and we make the inductive hypothesis that it is true when $i = j - 1$, so that

$$(\kappa_{j-1}', v_{j-1}') = (\kappa^*, g)(\kappa_{j-1}, v_{j-1}) = (\kappa^* \hat{g}(\kappa_{j-1}), g(v_{j-1})).$$

Since (κ_j, v_j) is adjacent to (κ_{j-1}, v_{j-1}) we have $\kappa_j = \kappa_{j-1} \phi(v_{j-1}, v_j)$, and the corresponding equation holds for the primed symbols as well. Thus:

$$\begin{aligned} \kappa_j' &= \kappa_{j-1}' \phi(v_{j-1}', v_j') = \kappa^* \hat{g}(\kappa_{j-1}) \phi(g(v_{j-1}), g(v_j)) \\ &= \kappa^* \hat{g}(\kappa_{j-1}) \hat{g}(\phi(v_{j-1}, v_j)) = \kappa^* \hat{g}(\kappa_{j-1} \phi(v_{j-1}, v_j)) \\ &= \kappa^* \hat{g}(\kappa_j). \end{aligned}$$

Consequently, (κ^*, g) takes (κ_j, v_j) to (κ_j', v_j'), and the result follows by the principle of induction. □

The requirement that a compatible K-chain should exist is rather restrictive. In fact, for a given graph Γ and group K, it is very likely that the only covering graph is the trivial one consisting of $|K|$ components each isomorphic with Γ. However, it is possible to choose K (depending on Γ) in such a way that a non-trivial covering graph always exists.

Let us suppose that a t-transitive graph Γ is given. We define K to be the free $\mathbb{Z}_2$-module on the set $E\Gamma$; thus, K is the direct product of $|E\Gamma|$ copies of $\mathbb{Z}_2$, and its elements are the formal products $\prod e_\alpha^{\sigma(\alpha)}$, where $\sigma(\alpha) = 0$ or 1 and the product is over all e_α in $E\Gamma$. The automorphism group $G = \text{Aut}(\Gamma)$ acts on K through its action on $E\Gamma$, and furthermore there is a K-chain ϕ on Γ defined by the rule $\phi(u,v) = e_i$, where $e_i = \{u,v\}$ regarded as an element of K. This K-chain is compatible with the actions of G on K and $S\Gamma$, and so the covering graph $\widetilde{\Gamma} = \widetilde{\Gamma}(K, \phi)$ exists and (by Proposition 19.4) its automorphism group is transitive on t-arcs. □

Theorem 19.5 *Let Γ be a t-transitive graph whose rank and co-rank are $r(\Gamma)$ and $s(\Gamma)$. Then, with the special choices of K and ϕ given above, the covering graph $\widetilde{\Gamma}$ consists of $2^{r(\Gamma)}$ connected components, each having $2^{s(\Gamma)}|V\Gamma|$ vertices.*

Proof Pick a vertex v of Γ, and let $\widetilde{\Gamma}_0$ denote the component of $\widetilde{\Gamma}$ which contains the vertex $(1,v)$. If

$$v = u_0, u_1, \ldots, u_l = v$$

are the vertices of a cycle in Γ, with edges $e_i = \{u_{i-1}, u_i\}$, then we have the following path in $\widetilde{\Gamma}_0$:

$$(1, v), \ (e_1, u_1), \ (e_1 e_2, u_2), \ \ldots, \ (e_1 e_2 \ldots e_l, v).$$

Conversely, the vertex (κ, v) is in $\widetilde{\Gamma}_0$ only if κ represents the edges of a cycle in Γ. Since there are $s(\Gamma)$ independent cycles in Γ, there are $2^{s(\Gamma)}$ elements κ in K such that (κ, v) is in $\widetilde{\Gamma}_0$. It follows that $\widetilde{\Gamma}_0$ has $2^{s(\Gamma)}|V\Gamma|$ vertices; further, $\widetilde{\Gamma}$ is vertex-transitive and so each component has this number of vertices. Finally, since

$$|V\widetilde{\Gamma}| = |K||V\Gamma| = 2^{|E\Gamma|}|V\Gamma| \quad \text{and} \quad r(\Gamma) + s(\Gamma) = |E\Gamma|,$$

there must be $2^{r(\Gamma)}$ components. □

Corollary 19.6 *There are infinitely many cubic 5-transitive graphs.*

Proof We know that there is at least one cubic 5-transitive graph, the graph Ω constructed at the end of the previous chapter. Applying the construction of Theorem 19.5 to Ω, we obtain a cubic 5-transitive graph $\widetilde{\Omega}_0$ with $2^{s(\Omega)}|V\Omega|$ vertices, and since $s(\Omega) > 0$ this graph is not isomorphic with Ω. We may repeat this process as often as we please, obtaining an infinite sequence of graphs with the required properties.

□

Of course the number of vertices used in Corollary 19.6 quickly becomes astronomical; for instance, the two graphs which follow Ω in the sequence have about 2^{21} and 2^{100000} vertices respectively. Biggs and Hoare (1983) have given an explicit construction for infinitely many cubic 5-transitive graphs which involves much smaller numbers (see **18b**).

Additional Results

19a *Double coverings*   Let G be the automorphism group of a connected graph Γ, and let G act on the group $\mathbb{Z}_2$ by the rule that $\hat{g}$ is the identity automorphism of $\mathbb{Z}_2$ for each g in G. Then the $\mathbb{Z}_2$-chain ϕ on Γ which assigns the non-identity element of $\mathbb{Z}_2$ to each side of Γ is compatible with the actions of G on $S\Gamma$ and $\mathbb{Z}_2$. The covering graph $\tilde{\Gamma}(\mathbb{Z}_2, \phi)$ is connected if and only if Γ is not bipartite. For example, applying the construction to the graph with 234 vertices described in **18a** we get a connected 5-transitive cubic graph with 468 vertices.

19b *The Desargues graph*   The construction of **19a** applied to Petersen's graph results in a cubic 3-transitive graph with 20 vertices. The vertices of this graph correspond to the points and lines in the Desargues configuration, with two vertices being adjacent if they correspond to an incident (point, line) pair. This graph was described by Coxeter (1950), together with several others derived from geometrical configurations.

19c *A threefold covering of $S(3)$*   The second 5-transitive cubic graph in order of magnitude is a graph with 90 vertices which is a threefold covering of the sextet graph $S(3)$ (see **18i**). Ito (1982) constructed an explicit $\mathbb{Z}_3$-chain on $S(3)$, which shows that the graph is a covering graph of $S(3)$ in the sense of this chapter.

19d *Another covering construction for 5-transitive cubic graphs*   Suppose that Γ is a cubic graph and $G = \text{Aut}(\Gamma)$ is a group of type 4^+. Then the automorphism $a^{-1}b$ fixes the vertices $\alpha_0, \alpha_1, \alpha_2$, and α_3 of the basic 4-arc $[\alpha]$, and (because the degree is 3) it must fix the other vertices β_1, β_2 adjacent to α_1, α_2 respectively. By considerations of order we see that this is the only non-identity automorphism with this property. So for each $e = \{v, w\} \in E\Gamma$ the group L_{vw} has order 2; in other words, there is a unique involution j_e which fixes e and the four vertices adjacent to e. The involutions j_e generate the group G^* (Proposition 18.2) which is normal, of index 1 or 2 in G. Consequently G acts by conjugation as a group of automorphisms of G^*.

If we take $K = G^*$ and define a K-chain on Γ by

$$\phi(v, w) = j_e,$$

then the compatibility condition is satisfied and, by Proposition 19.4, we have a graph $\tilde{\Gamma}$ on which $K \hat{\times} G$ acts 4-transitively. However, there is a bonus. As shown by Biggs (1982b) there is an 'extra' automorphism $\tilde{\Gamma}$, so that $\tilde{\Gamma}$ is in fact a 5-transitive graph.

19e *A 5-transitive cubic graph with* 2352 *vertices*   The simplest case of **19d** is when $\Gamma = S(7)$, a graph with 14 vertices also known as the *Heawood graph*. In this case both Γ and its 5-transitive covering graph $\widetilde{\Gamma}$ with 2352 vertices can be constructed directly in terms of the seven-point plane $PG(2,2)$ (Biggs 1982a).

19f *Conway generators for the covering graph*   Let a be the Conway generator for the t-transitive group G of Γ, with respect to the t-arc $[\alpha]$, and suppose ϕ is a compatible K-chain. Then the corresponding generator $\tilde{a}$ for the group $K \hat{\times} G$ of $\widetilde{\Gamma}(K, \phi)$ is (λ, a), where $\lambda = \phi(\alpha_0, \alpha_1)$.

19g *Homological coverings*   Let Γ be a graph with co-rank s and let R be a ring. The 'first homology group with coefficients in R' of a graph Γ is the direct product R^s of s copies of R. (This a just a mild generalization of the cycle space defined in Chapter 4.) The functorial properties of homology imply that the automorphism group of Γ acts as a group of automorphisms of the homology group, and so a covering graph $\widetilde{\Gamma}$ can be constructed using $K = R^s$. Biggs (1984b) gave an explicit form of this construction and showed that when $R = \mathbb{Z}$ the number of components of $\widetilde{\Gamma}$ is equal to the tree-number of Γ.

19h *The Pappus graph*   In the homological covering construction, take $R = \mathbb{Z}_3$ as the coefficient group and $\Gamma = K_{3,3}$, so that $s = 4$ and $K = \mathbb{Z}_3^4$. The covering graph in this case has 27 components, each with $6 \times 3^4/27 = 18$ vertices. Each component is a copy of the *Pappus graph*, whose vertices correspond to the points and lines of the Pappus configuration, with adjacent vertices corresponding to an incident (point, line) pair. See also Coxeter (1950).

20

Distance-transitive graphs

In Chapter 15 a connected graph Γ was defined to be distance-transitive if, for any vertices u, v, x, y of Γ satisfying $\partial(u, v) = \partial(x, y)$, there is an automorphism g of Γ which takes u to x and v to y.

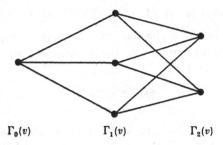

$\Gamma_0(v)$ $\Gamma_1(v)$ $\Gamma_2(v)$

Figure 16: a distance-partition of $K_{3,3}$.

It is helpful to recast the definition. For any vertex v of a connected graph Γ we define

$$\Gamma_i(v) = \{u \in V\Gamma \mid \partial(u, v) = i\},$$

where i is a non-negative integer not exceeding d, the diameter of Γ. It is clear that $\Gamma_0(v) = \{v\}$, and $V\Gamma$ is partitioned into the disjoint subsets $\Gamma_0(v), \ldots, \Gamma_d(v)$, for each v in $V\Gamma$. Small graphs may be depicted in a manner which emphasises this partition by arranging their vertices in columns, according to distance from an arbitrary vertex v. For example, $K_{3,3}$ is displayed in this way in Figure 16.

Lemma 20.1 *A connected graph Γ with diameter d and automorphism group $G = \mathrm{Aut}(\Gamma)$ is distance-transitive if and only if it is vertex-transitive and the vertex-stabilizer G_v is transitive on the set $\Gamma_i(v)$, for each $i \in \{0, 1, \ldots, d\}$, and each $v \in V\Gamma$.*

Proof Suppose that Γ is distance-transitive. Taking $u = v$ and $x = y$ in the definition (as given above), we see that Γ is vertex-transitive. Taking $y = v$, we see that G_v is transitive on $\Gamma_i(v)$ $(0 \le i \le d)$.

Conversely, suppose vertices u, v, x, y are given such that $\partial(u, v) = \partial(x, y) = i$. Let g be an automorphism such that $g(v) = y$ and let $h \in G_y$ be such that $h(g(u)) = x$. Then hg takes u to x and v to y.

$\square$

As we shall see, the adjacency algebra (defined in Chapter 2) plays a major part in the study of distance-transitive graphs. In preparation for the algebraic theory we begin by investigating some simple combinatorial consequences of the definition.

For any connected graph Γ, any vertices u, v of Γ, and any non-negative integers h and i, define $s_{hi}(u, v)$ to be the number of vertices of Γ whose distance from u is h and whose distance from v is i. That is,

$$s_{hi}(u, v) = |\{w \in V\Gamma \mid \partial(u, w) = h \text{ and } \partial(v, w) = i\}|.$$

In a distance-transitive graph the numbers $s_{hi}(u, v)$ depend, not on the individual pair (u, v), but only on the distance $\partial(u, v)$. So if $\partial(u, v) = j$ we shall write

$$s_{hij} = s_{hi}(u, v).$$

Definition 20.2 The *intersection numbers* of a distance-transitive graph with diameter d are the numbers s_{hij}, where h, i and j belong to the set $\{0, 1, \ldots, d\}$.

Clearly there are $(d + 1)^3$ intersection numbers, but it turns out that there are many identities relating them; and in due course we shall show that just $2d$ of them are sufficient to determine the rest.

Consider the intersection numbers with $h = 1$. For a fixed j, s_{1ij} is the number of vertices w such that w is adjacent to u and $\partial(v, w) = i$, when $\partial(u, v) = j$. Now, if w is adjacent to u and $\partial(u, v) = j$, then $\partial(v, w)$ must be one of the numbers $j - 1, j, j + 1$; in other words

$$s_{1ij} = 0 \quad \text{if} \quad i \ne j - 1, j, j + 1.$$

More generally, $s_{hij} = 0$ if the largest of h, i, j is greater than the sum of the other two.

For the intersection numbers s_{1ij} which are not identically zero we shall use the notation

$$c_j = s_{1,j-1,j} \quad a_j = s_{1,j,j}, \quad b_j = s_{1,j+1,j},$$

where $0 \le j \le d$, and it is convenient to leave c_0 and b_d undefined. The numbers c_j, a_j, b_j have the following simple interpretation in terms of the diagrammatic representation of Γ introduced at the beginning of this chapter. If we pick an arbitrary vertex v and a vertex u in $\Gamma_j(v)$, then u is adjacent to c_j vertices in $\Gamma_{j-1}(v)$, a_j vertices in $\Gamma_j(v)$, and b_j vertices in $\Gamma_{j+1}(v)$. These numbers are independent of u and v, provided that $\partial(u,v) = j$.

Definition 20.3   The *intersection array* of a distance-transitive graph is

$$\iota(\Gamma) = \left\{ \begin{matrix} * & c_1 & \cdots & c_j & \cdots & c_d \\ a_0 & a_1 & \cdots & a_j & \cdots & a_d \\ b_0 & b_1 & \cdots & b_j & \cdots & * \end{matrix} \right\}.$$

For example, consider the cube Q_3, which is a distance-transitive graph with diameter 3. From the representation in Figure 17 we may write down its intersection array.

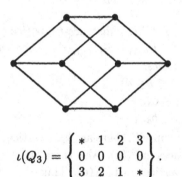

$$\iota(Q_3) = \left\{ \begin{matrix} * & 1 & 2 & 3 \\ 0 & 0 & 0 & 0 \\ 3 & 2 & 1 & * \end{matrix} \right\}.$$

Figure 17: Q_3 as a distance-transitive graph.

We observe that a distance-transitive graph is vertex-transitive, and consequently regular, of degree k say. Clearly we have $b_0 = k$ and $a_0 = 0, c_1 = 1$. Further, since each column of the intersection array sums to k, if we are given the first and third rows we can calculate the middle row. Thus it is both logically sufficient and typographically convenient to use the alternative notation

$$\iota(\Gamma) = \{k, b_1, \ldots, b_{d-1}; 1, c_2, \ldots, c_d\}.$$

However, the original notation of Definition 20.3 is intuitively helpful, and we shall continue to use it whenever it seems appropriate. In due course we shall see that the intersection array determines all the intersection numbers s_{hij}.

Many well-known families of graphs are distance-transitive, although this apparent profusion of examples is rather misleading, because the property is, in some senses, very rare. The complete graphs K_n and the complete bipartite graphs $K_{k,k}$ are distance-transitive. Their diameters are 1 and 2 respectively, and the intersection arrays are:

$$\iota(K_n) = \left\{ \begin{array}{ccc} * & 1 \\ 0 & n-2 \\ n-1 & * \end{array} \right\}, \quad \iota(K_{k,k}) = \left\{ \begin{array}{ccc} * & 1 & k \\ 0 & 0 & 0 \\ k & k-1 & * \end{array} \right\}.$$

The triangle graphs $\Delta_t = L(K_t)$ (p. 21) are distance-transitive, with diameter 2, and for $t \geq 4$:

$$\iota(\Delta_t) = \left\{ \begin{array}{ccc} * & 1 & 4 \\ 0 & t-2 & 2t-8 \\ 2t-4 & t-3 & * \end{array} \right\}.$$

Many other distance-transitive graphs will be described in the following chapters.

Denote by k_i ($0 \leq i \leq d$) the number of vertices in $\Gamma_i(v)$ for any vertex v; in particular $k_0 = 1$ and $k_1 = k$.

Proposition 20.4 *Let Γ be a distance-transitive graph whose intersection array is $\{k, b_1, \ldots, b_{d-1}; 1, c_2, \ldots, c_d\}$. Then we have the following equations and inequalities:*

(1) $k_{i-1}b_{i-1} = k_i c_i$ $(1 \leq i \leq d)$.

(2) $1 \leq c_2 \leq c_3 \leq \ldots \leq c_d$.

(3) $k \geq b_1 \geq b_2 \geq \ldots \geq b_{d-1}$.

Proof (1) For any v in $V\Gamma$, there are k_{i-1} vertices in $\Gamma_{i-1}(v)$ and each is joined to b_{i-1} vertices in $\Gamma_i(v)$. Also there are k_i vertices in $\Gamma_i(v)$ and each is joined to c_i vertices in $\Gamma_{i-1}(v)$. Thus the number of edges with one end in $\Gamma_{i-1}(v)$ and one end in $\Gamma_i(v)$ is $k_{i-1}b_{i-1} = k_i c_i$.

(2) Suppose u is in $\Gamma_{i+1}(v)$ ($1 \leq i \leq d-1$). Pick a path $v, x, \ldots, u$ of length $i+1$; then $\partial(x, u) = i$. If w is in $\Gamma_{i-1}(x) \cap \Gamma_1(u)$, then $\partial(v, w) = i$, and so w is in $\Gamma_i(v) \cap \Gamma_1(u)$. It follows that

$$c_i = |\Gamma_{i-1}(x) \cap \Gamma_1(u)| \leq |\Gamma_i(v) \cap \Gamma_1(u)| = c_{i+1}.$$

(3) This is proved by an argument analogous to that used in (2). $\square$

Proposition 20.4 provides some simple constraints which must be satisfied if an arbitrary array is to be the intersection array of some distance-transitive graph. We shall obtain much more restrictive conditions in the next chapter. However, in order to derive these conditions, we need not postulate that the graph is distance-transitive, but merely that it has the combinatorial regularity implied by the existence of an intersection array. This is the justification for the following definition.

Definition 20.5 A *distance-regular graph* is a regular connected graph with degree k and diameter d, for which following holds. There are natural numbers

$$b_0 = k, \quad b_1, \ldots, b_{d-1}, \quad c_1 = 1, \quad c_2, \ldots, c_d,$$

such that for each pair (u, v) of vertices satisfying $\partial(u, v) = j$ we have

(1) the number of vertices in $\Gamma_{j-1}(v)$ adjacent to u is c_j $(1 \leq j \leq d)$;

(2) the number of vertices in $\Gamma_{j+1}(v)$ adjacent to u is b_j $(0 \leq j \leq d-1)$.

The array $\{k, b_1, \ldots, b_{d-1}; 1, c_2, \ldots, c_d\}$ is the *intersection array* of Γ.

Note that a distance-regular graph with diameter $d = 2$ is simply a *strongly regular* graph, as defined in **3c**. In terms of the general definition, the parameters a and c of a strongly regular graph are given by $a = k - 1 - b_1$ and $c = c_2$.

It is clear that a distance-transitive graph is distance-regular, but the converse is not true. Although many 'familiar' examples of distance-regular graphs are distance-transitive, it is possible to construct arbitrarily large families of distance-regular graphs which are not distance-transitive. Several examples will be given in the course of the following chapters.

We shall now construct a basis for the adjacency algebra of a distance-regular graph. Given a graph Γ with vertex-set $\{v_1, \ldots, v_n\}$ and diameter d, define a set $\{A_0, A_1, \ldots, A_d\}$ of $n \times n$ *distance matrices* as follows:

$$(A_h)_{rs} = \begin{cases} 1 & \text{if } \partial(v_r, v_s) = h; \\ 0 & \text{otherwise.} \end{cases}$$

In particular $A_0 = I$, and A_1 is the usual adjacency matrix A of Γ. We notice that $A_0 + A_1 + \ldots + A_d = J$, where J is the all-1 matrix.

Lemma 20.6 *Let Γ be a distance-regular graph and let*

$$\{k, b_1, \ldots, b_{d-1}; 1, c_2, \ldots, c_d\}$$

be its intersection array. For $1 \leq i \leq d-1$, define $a_i = k - b_i - c_i$; then

$$AA_i = b_{i-1}A_{i-1} + a_iA_i + c_{i+1}A_{i+1} \quad (1 \leq i \leq d-1).$$

Proof From the definition of A and A_i it follows that $(AA_i)_{rs}$ is the number of vertices w of Γ such that $\partial(v_r, w) = 1$ and $\partial(v_s, w) = i$. If there are any such vertices w, then $\partial(v_r, v_s)$ must be one of the numbers $i - 1, i, i + 1$, and the number of vertices w in these three cases is b_{i-1}, a_i, c_{i+1}, respectively. Thus $(\mathbf{AA}_i)_{rs}$ is equal to the (r,s)-entry of the matrix on the right-hand side. □

Theorem 20.7 (Damerell 1973)   *Let Γ be a distance-regular graph with diameter d. Then $\{\mathbf{A}_0, \mathbf{A}_1, \ldots, \mathbf{A}_d\}$ is a basis for the adjacency algebra $\mathcal{A}(\Gamma)$, and consequently the dimension of $\mathcal{A}(\Gamma)$ is $d+1$.*

Proof   By recursive applications of the lemma we see that $\mathbf{A}_i$ is a polynomial $p_i(\mathbf{A})$, for $i = 2, \ldots, d$. The form of the recursion shows that the degree of p_i is at most i, and since $\mathbf{A}_0, \mathbf{A}_1, \ldots, \mathbf{A}_d$ are linearly independent (exactly one of them has a non-zero entry in any given position) the degree of p_i is exactly i.

Since $\mathbf{A}_0 + \mathbf{A}_1 + \ldots + \mathbf{A}_d = \mathbf{J}$ and Γ is k-regular we have

$$(\mathbf{A} - k\mathbf{I})(\mathbf{A}_0 + \mathbf{A}_1 + \ldots + \mathbf{A}_d) = \mathbf{0}.$$

The left-hand side is a polynomial in $\mathbf{A}$ of degree $d+1$, so the dimension of $\mathcal{A}(\Gamma)$ is at most $d+1$. However, since $\{\mathbf{A}_0, \mathbf{A}_1, \ldots, \mathbf{A}_d\}$ is a set of $d+1$ linearly independent members of $\mathcal{A}(\Gamma)$, it is a basis, and the dimension is equal to $d+1$. □

It follows from Theorem 20.7 that a distance-regular graph has just $d+1$ distinct eigenvalues, the minimum number possible for a graph of diameter d. These eigenvalues, and a remarkable formula for calculating their multiplicities, form the subject of the next chapter.

The full set of $(d+1)^3$ intersection numbers can be defined for a distance-regular graph; this is a trivial remark for a distance-transitive graph, but it requires proof in the distance-regular case. In the course of the proof we shall relate these intersection numbers to the basis $\{\mathbf{A}_0, \mathbf{A}_1, \ldots, \mathbf{A}_d\}$ of $\mathcal{A}(\Gamma)$.

Proposition 20.8   *Let Γ be a distance-regular graph with diameter d.*
(1) *The numbers $s_{ih}(u,v)$, $h, i \in \{0, 1, \ldots, d\}$, depend only on $\partial(u,v)$.*
(2) *If $s_{hi}(u,v) = s_{hij}$ when $\partial(u,v) = j$, then*

$$\mathbf{A}_h \mathbf{A}_i = \sum_{j=0}^{d} s_{hij} \mathbf{A}_j.$$

Proof   We prove both parts in one argument. Since $\{\mathbf{A}_0, \mathbf{A}_1, \ldots, \mathbf{A}_d\}$ is a basis for $\mathcal{A}(\Gamma)$, the product $\mathbf{A}_h \mathbf{A}_i$ is a linear combination $\sum t_{hij} \mathbf{A}_j$. Now

$$(\mathbf{A}_h \mathbf{A}_i)_{rs} = s_{hi}(v_r, v_s),$$

and there is just one member of the basis whose (r,s)-entry is 1: it is that $\mathbf{A}_j$ for which $\partial(v_r, v_s) = j$. Thus $s_{hi}(v_r, v_s) = t_{hij}$, and so

$s_{hi}(v_r, v_s)$ depends only on $\partial(v_r, v_s)$. Further, the coefficient t_{hij} is just the intersection number s_{hij}. ☐

At this point a few historical remarks are in order. The theory which underlies our treatment of the adjacency algebra of a distance-regular graph was developed in two quite different contexts. First, the association schemes used by Bose in the statistical design of experiments led to an association algebra (Bose and Mesner 1959), which corresponds to our adjacency algebra. Bose and others also studied strongly regular graphs which, as we have noted, are just distance-regular graphs with diameter 2. Secondly, the work of Schur (1933) and Wielandt (1964) on the commuting algebra, or centralizer ring, of a permutation group, culminated in the paper of Higman (1967) which employs graph-theoretic ideas very closely related to those of this chapter. The discovery of sporadic simple groups as the automorphism groups of strongly regular graphs (for example by Higman and Sims (1968)) gave a powerful impetus to work in this area. The formulation in terms of the properties of distance-transitivity and distance-regularity was developed by the present author and some of his colleagues in the years 1969–1973, and a consolidated account appeared in the first edition of this book (1974). In the last twenty years an extensive literature has been accumulating. The reader is referred to the now-standard text of Brouwer, Cohen and Neumaier [BCN], which admirably covers the state of the art up to 1989, and contains a bibliography of 800 items.

Additional Results

20a *The cube graphs* The k-cube, Q_k, is the graph defined as follows: the vertices of Q_k are the 2^k symbols $(\epsilon_1, \epsilon_2, \dots, \epsilon_k)$, where $\epsilon_i = 0$ or 1 $(1 \leq i \leq k)$, and two vertices are adjacent when the symbols differ in exactly one coordinate. The graph Q_k $(k \geq 2)$ is distance-transitive, with degree k and diameter k, and the intersection array is

$$\{k, k-1, k-2, \dots, 1; 1, 2, 3, \dots, k\}.$$

20b *The odd graphs yet again* The odd graphs O_k $(k \geq 2)$ are distance-transitive, with degree k and diameter $k-1$. The intersection array, in the cases $k = 2l - 1$ and $k = 2l$ respectively, is

$$\{2l-1, 2l-2, 2l-2, \dots, l+1, l+1, l; \ 1, 1, 2, 2, \dots, l-1, l-1\},$$

$$\{2l, 2l-1, 2l-1, \dots, l+1, l+1; \ 1, 1, 2, 2, \dots, l-1, l-1, l\}.$$

20c *A distance-regular graph which is not distance-transitive*   Let Ψ denote the graph whose vertices are the 26 symbols a_i, b_i (where i is an integer modulo 13), and in which:

$$a_i \text{ and } a_j \text{ are adjacent } \Leftrightarrow |i - j| = 1, 3, 4;$$
$$b_i \text{ and } b_j \text{ are adjacent } \Leftrightarrow |i - j| = 2, 5, 6;$$
$$a_i \text{ and } b_j \text{ are adjacent } \Leftrightarrow i - j = 0, 1, 3, 9.$$

Then Ψ is distance-regular, with diameter 2, and its intersection array is $\{10, 6; 1, 4\}$. But Ψ is not distance-transitive; in fact there is no automorphism taking a vertex a_i to a vertex b_j (Adel'son-Velskii *et al.* 1969).

20d *Strengthening the distance-transitivity condition*   A connected simple graph is *r-ply transitive* if, for any two ordered r-tuples of vertices $(x_1, \ldots, x_r)$ and $(y_1, \ldots, y_r)$ satisfying $\partial(x_i, x_j) = \partial(y_i, y_j)$ for all i, j, there is an automorphism g for which $g(x_i) = y_i$ ($1 \leq i \leq r$). Clearly, a 1-ply transitive graph is vertex-transitive, and a 2-ply transitive graph is distance-transitive. Meredith (1976) showed that the only 3-ply transitive graphs with girth greater than 4 (equivalently $c_2 = 1$) are the cycles.

20e *6-ply transitive graphs* (Cameron 1980)   The following is a complete list of all 6-ply transitive graphs.
(i) The complete multipartite graphs with parts of equal size (including the complete graphs as the case when the parts have size 1).
(ii) The complete bipartite graphs with the edges of a complete matching deleted.
(iii) The cycles.
(iv) $L(K_{3,3})$.
(v) The icosahedron.
(vi) The graph whose vertices are the 3-subsets of a 6-set, two vertices being adjacent whenever they have two common members.

20f *Strongly regular graphs and partial geometries*   A *partial geometry* $\text{pg}(s, t, \alpha)$ is an incidence structure of points and lines such that every line has $s + 1$ points, every point is on $t + 1$ lines, two distinct lines meet in at most one point, and for every non-incident (point, line) pair (p, l) there are α lines through p that meet l. The graph whose vertices are the points, two being adjacent if they are collinear, is strongly regular with parameters $k = s(t + 1)$, $a = t(\alpha - 1) + s - 1$, $c = \alpha(t + 1)$. Equivalently, it is a distance-regular graph with intersection array

$$\{s(t + 1), (s - \alpha + 1)t;\ 1, \alpha(t + 1)\}.$$

20g *Symmetric designs as distance-regular graphs* A *symmetric design* with parameters (v, k, λ) is a set P of points and a set B of blocks such that $|P| = |B| = v$, each block has k points and each point is in k blocks, and each pair of points is in λ blocks. It follows from the definition that $(v-1)\lambda = k(k-1)$. When $\lambda = 1$ a symmetric design is called a *projective plane*.

The graph whose vertices are the points and blocks of a symmetric design, two being adjacent when they are incident, is distance-regular with intersection array

$$\{k, k - 1, k - \lambda; \ 1, \lambda, k\}.$$

For example, when $\lambda = 1$ we have the incidence graph of a projective plane; the case $k = 3$ is Heawood's graph $S(7)$ mentioned in **18h**. If the projective plane is Desarguesian (that is, if it can be coordinatized using a finite field) then the corresponding graph is distance-transitive.

20h *The classification problem for DT and DR graphs* For each $k \geq 3$ there are only finitely many DT graphs with degree k. This has been proved in several ways: see Cameron (1982) and Weiss (1985), for example. For DR graphs the result has been established only in the case $k = 3$ (Biggs, Boshier and Shawe-Taylor (1986); see **21i**).

For the general DR case the problem is to find an upper bound for the diameter d in terms of k. Such a result could be regarded as a strengthening of the monotonicity conditions (2) and (3) of Proposition 20.4, in which we seek to bound the number of repeated values among the columns (c_i, a_i, b_i) of the intersection array. An important result on these lines was obtained by Ivanov (1983).

21

Feasibility of intersection arrays

In this chapter we shall study the following question. Suppose that an arbitrary array of integers $\{k, b_1, \ldots, b_{d-1}; 1, c_2, \ldots, c_d\}$ is given: when is there a distance-regular graph with this as its intersection array?

The results obtained in the previous chapter provide some simple necessary conditions. For example, part (1) of Proposition 20.4 yields an explicit formula for the numbers $k_i = |\Gamma_i(v)|$:

$$k_i = (kb_1 \ldots b_{i-1})/(c_2 c_3 \ldots c_i) \quad (2 \le i \le d).$$

These numbers must be integers, so we have a non-trivial constraint on the intersection array. Similarly, the monotonicity conditions in parts (2) and (3) of Proposition 20.4 must be satisfied.

There are also some elementary parity conditions. Let $n = 1 + k_1 + \ldots + k_d$ be the number of vertices of the putative graph, then if k is odd, n must be even. That is, $nk \equiv 0 \pmod 2$. Similarly, considering the induced subgraph defined by the vertices in $\Gamma_i(v)$, we see that $k_i a_i \equiv 0 \pmod 2$ for $1 \le i \le d$, where $a_i = k - b_i - c_i$.

These conditions are quite restrictive, yet they are satisfied by many arrays which are not realised by any graph. For example, $\{3, 2, 1; 1, 1, 3\}$ passes all these tests, and would represent a graph with degree 3, diameter 3, and 12 vertices. In this case, simple (but special) arguments can be used to prove that there is no graph. The main result of this chapter is a general condition which rules out a multitude of examples of this kind.

Recall that the adjacency algebra $\mathcal{A}(\Gamma)$ of a distance-regular graph Γ

has as a basis the $d+1$ distance matrices $\{\mathbf{A}_0, \mathbf{A}_1, \ldots, \mathbf{A}_d\}$, which satisfy $\mathbf{A}_h \mathbf{A}_i = \sum s_{hij} \mathbf{A}_j$. This equation can be interpreted as saying that left-multiplication by $\mathbf{A}_h$, regarded as a linear mapping of $\mathcal{A}(\Gamma)$ with respect to the given basis, is faithfully represented by the $(d+1) \times (d+1)$ matrix $\mathbf{B}_h$ defined by

$$(\mathbf{B}_h)_{ij} = s_{hij}.$$

(This representation seems natural for our purposes although it is the transpose of the one most commonly employed. Since the algebra $\mathcal{A}(\Gamma)$ is commutative, the difference is immaterial.) The existence of this representation is sufficiently important to justify a formal statement.

Proposition 21.1 *The adjacency algebra $\mathcal{A}(\Gamma)$ of a distance-regular graph Γ with diameter d can be faithfully represented by an algebra of matrices with $d+1$ rows and columns. A basis for this representation is the set $\{\mathbf{B}_0, \mathbf{B}_1, \ldots, \mathbf{B}_d\}$, where $(\mathbf{B}_h)_{ij}$ is the intersection number s_{hij} for $h, i, j \in \{0, 1, \ldots, d\}$.* ☐

The members of $\mathcal{A}(\Gamma)$ can now be regarded as square matrices of size $d + 1$ (instead of n), a considerable simplification. What is more, the matrix $\mathbf{B}_1$ alone is sufficient. To see this, we notice first that, since $(\mathbf{B}_1)_{ij} = s_{1ij}$, the matrix $\mathbf{B}_1$ is *tridiagonal*:

$$\mathbf{B}_1 = \begin{bmatrix} 0 & 1 & & & & & \\ k & a_1 & c_2 & & & & \\ & b_1 & a_2 & \cdot & & & \\ & & b_2 & \cdot & \cdot & & \\ & & & \cdot & \cdot & \cdot & \\ & & & & \cdot & \cdot & c_d \\ & & & & & \cdot & a_d \end{bmatrix}.$$

We shall often write $\mathbf{B}$ for $\mathbf{B}_1$, and refer to $\mathbf{B}$ as the *intersection matrix* of Γ. Note that it is just another way of writing the intersection array. Now, since the matrices $\mathbf{B}_i$ are images of the matrices $\mathbf{A}_i$ under a faithful representation, the equation obtained in Lemma 20.6 carries over:

$$\mathbf{B}\mathbf{B}_i = b_{i-1}\mathbf{B}_{i-1} + a_i\mathbf{B}_i + c_{i+1}\mathbf{B}_{i+1} \quad (1 \leq i \leq d-1).$$

Consequently each $\mathbf{B}_i$ is a polynomial in $\mathbf{B}$ with coefficients which depend only on the entries of $\mathbf{B}$. It follows from this (in theory) that $\mathcal{A}(\Gamma)$ and the spectrum of Γ are determined by $\mathbf{B}$, which in turn is determined by the intersection array $\iota(\Gamma)$. We shall now give an explicit demonstration of this fact.

Proposition 21.2 *Let Γ be a distance-regular graph with degree k and diameter d. Then Γ has $d + 1$ distinct eigenvalues $k = \lambda_0, \lambda_1, \ldots, \lambda_d$, which are the eigenvalues of the intersection matrix $\mathbf{B}$.*

Proof We noted in Chapter 20 that Γ has exactly $d+1$ distinct eigenvalues. Since $\mathbf{B}$ is the image of the adjacency matrix $\mathbf{A}$ under a faithful representation, the minimum polynomials of $\mathbf{A}$ and $\mathbf{B}$ coincide, and so the eigenvalues of $\mathbf{A}$ are the same as those of $\mathbf{B}$. $\square$

Each eigenvalue λ, common to $\mathbf{A}$ and $\mathbf{B}$, is a simple eigenvalue of $\mathbf{B}$, since $\mathbf{B}$ is a matrix of size $d+1$. However, the multiplicity $m(\lambda)$ of λ *as an eigenvalue of* $\mathbf{A}$ will usually be greater than one, since the sum of the multiplicities is n, the number of vertices. We shall show how $m(\lambda)$ can be calculated from $\mathbf{B}$ alone.

Let us regard λ as an indeterminate, and define a sequence of polynomials in λ, with rational coefficients, by the recursion:

$$v_0(\lambda) = 1, \quad v_1(\lambda) = \lambda,$$

$$c_{i+1}v_{i+1}(\lambda) + (a_i - \lambda)v_i(\lambda) + b_{i-1}v_{i-1}(\lambda) = 0 \quad (i = 1, 2, \ldots, d-1).$$

The polynomial $v_i(\lambda)$ has degree i in λ, and comparing the definition with Lemma 20.6 we see that

$$\mathbf{A}_i = v_i(\mathbf{A}) \quad (i = 0, 1, \ldots, d).$$

Another interpretation of the sequence $\{v_i(\lambda)\}$ is as follows. If we introduce the column vector $\mathbf{v}(\lambda) = [v_0(\lambda), v_1(\lambda), \ldots, v_d(\lambda)]^t$, then the defining equations are those which arise when we put $v_0(\lambda) = 1$ and solve the system $\mathbf{B}\mathbf{v}(\lambda) = \lambda\mathbf{v}(\lambda)$, using one row of $\mathbf{B}$ at a time, and stopping at row $d-1$. The last row of $\mathbf{B}$ gives rise to an equation representing the condition that $\mathbf{v}(\lambda)$ is an eigenvector of $\mathbf{B}$ corresponding to the eigenvalue of λ. The roots of this equation in λ are the eigenvalues $\lambda_0, \lambda_1, \ldots, \lambda_d$ of $\mathbf{B}$, and so a *right* eigenvector $\mathbf{v}_i$ corresponding to λ_i has components $(\mathbf{v}_i)_j = v_j(\lambda_i)$.

It is convenient to consider also the *left* eigenvector $\mathbf{u}_i$ corresponding to λ_i; this is a row vector satisfying $\mathbf{u}_i\mathbf{B} = \lambda_i\mathbf{u}_i$. We shall say that a vector $\mathbf{x}$ is *standard* when $x_0 = 1$.

Lemma 21.3 *Suppose that $\mathbf{u}_i$ and $\mathbf{v}_i$ are standard left and right eigenvectors corresponding to the eigenvalue λ_i of $\mathbf{B}$. Then $(\mathbf{v}_i)_j = k_j(\mathbf{u}_i)_j$, for all $i, j \in \{0, 1, \ldots, d\}$.*

Proof Each eigenvalue of $\mathbf{B}$ is simple, and so there is a one-dimensional space of corresponding eigenvectors. It follows that there are unique standard eigenvectors $\mathbf{u}_i$ and $\mathbf{v}_i$. (If $(\mathbf{u}_i)_0$ or $(\mathbf{v}_i)_0$ were zero, then the tridiagonal form of $\mathbf{B}$ would imply that $\mathbf{u}_i = \mathbf{0}, \mathbf{v}_i = \mathbf{0}$.)

Let $\mathbf{K}$ denote the diagonal matrix with diagonal entries $k_0, k_1, \ldots, k_d$. Using the equations $b_{i-1}k_{i-1} = c_ik_i$ ($2 \leq i \leq d$) we may check that $\mathbf{B}\mathbf{K}$

is a symmetric matrix; that is,

$$\mathbf{BK} = (\mathbf{BK})^t = \mathbf{KB}^t.$$

Thus, if $\mathbf{u}_i\mathbf{B} = \lambda_i\mathbf{u}_i$ $(0 \le i \le d)$, we have

$$\mathbf{BKu}_i^t = \mathbf{KB}^t\mathbf{u}_i^t = \mathbf{K}(\mathbf{u}_i\mathbf{B})^t = \mathbf{K}(\lambda_i\mathbf{u}_i)^t = \lambda_i\mathbf{Ku}_i^t.$$

In other words, $\mathbf{Ku}_i^t$ is a right eigenvector of $\mathbf{B}$ corresponding to λ_i. Also $(\mathbf{Ku}_i^t)_0 = 1$, and so by the uniqueness of $\mathbf{v}_i$, it follows that $\mathbf{Ku}_i^t = \mathbf{v}_i$. □

We notice that, when $i \ne l$, the inner product $(\mathbf{u}_i, \mathbf{v}_l)$ is zero, since

$$\lambda_i(\mathbf{u}_i, \mathbf{v}_l) = \mathbf{u}_i\mathbf{Bv}_l = \lambda_l(\mathbf{u}_i, \mathbf{v}_l).$$

Our main result is that the inner product with $i = l$ determines the multiplicity $m(\lambda_i)$.

Theorem 21.4 *With the notation above, the multiplicity of the eigenvalue λ_i of a distance-regular graph with n vertices is*

$$m(\lambda_i) = \frac{n}{(\mathbf{u}_i, \mathbf{v}_i)} \quad (0 \le i \le d).$$

Proof For $i = 0, 1, \ldots, d$ define

$$\mathbf{L}_i = \sum_{j=0}^{d}(\mathbf{u}_i)_j\mathbf{A}_j.$$

We can calculate the trace of $\mathbf{L}_i$ in two ways. First, the trace of $\mathbf{A}_j$ is zero $(j \ne 0)$, and $\mathbf{A}_0 = \mathbf{I}$, so that

$$\text{tr}(\mathbf{L}_i) = (\mathbf{u}_i)_0\,\text{tr}(\mathbf{I}) = n.$$

On the other hand, since $\mathbf{A}_j = v_j(\mathbf{A})$, the eigenvalues of $\mathbf{A}_j$ are $v_j(\lambda_0)$, $\ldots, v_j(\lambda_d)$, with multiplicities $m(\lambda_0), \ldots, m(\lambda_d)$; consequently the trace of $\mathbf{A}_j$ is $\sum m(\lambda_l)v_j(\lambda_l)$. Thus

$$\text{tr}(\mathbf{L}_i) = \sum_j(\mathbf{u}_i)_j \sum_l m(\lambda_l)(\mathbf{v}_l)_j$$
$$= \sum_l m(\lambda_l)(\mathbf{u}_i, \mathbf{v}_l)$$
$$= m(\lambda_i)(\mathbf{u}_i, \mathbf{v}_i),$$

which gives the required result. □

In the context of our question about the realisability of a given array, we shall view Theorem 21.4 in the following way. The numbers $n/(\mathbf{u}_i, \mathbf{v}_i)$, which are completely determined by the array, represent multiplicities of the eigenvalues of the adjacency matrix of a supposed graph, and consequently if there is such a graph, they must be positive integers. This turns out to be a very powerful condition.

Definition 21.5   The array $\{k, b_1, \ldots, b_{d-1}; 1, c_2, \ldots, c_d\}$ is *feasible* if the following conditions are satisfied.

(1) The numbers $k_i = (kb_1 \ldots b_{i-1})/(c_2 c_3 \ldots c_i)$ are integers $(2 \le i \le d)$.

(2) $k \ge b_1 \ge \ldots \ge b_{d-1}$ and $1 \le c_2 \le \ldots \le c_d$.

(3) If $n = 1 + k + k_2 + \ldots + k_d$ and $a_i = k - b_i - c_i$ $(1 \le i \le d-1)$, $a_d = k - c_d$, then $nk \equiv 0 \pmod 2$ and $k_i a_i \equiv 0 \pmod 2$.

(4) The numbers $n/(\mathbf{u}_i, \mathbf{v}_i)$ are positive integers $(0 \le i \le d)$.

It should be noted that the definition of feasibility given above is a matter of convention. The conditions stated are not sufficient for the existence of a graph with the given array, and indeed there are many other, independent, 'feasibility' conditions. Some useful ones are given in **21c**, **21d**, and **21e**; the standard reference [BCN] provides a comprehensive treatment. The four conditions which comprise our definition of feasibility are chosen because they are particularly useful, and any reasonable way of testing a given array will surely include them.

The four conditions are easy to apply in practice. The calculation of $n/(\mathbf{u}_i, \mathbf{v}_i)$ is facilitated by Lemma 21.3, which implies that

$$(\mathbf{u}_i, \mathbf{v}_i) = \sum k_j (\mathbf{u}_i)_j^2 = \sum \frac{(\mathbf{v}_i)_j^2}{k_j}.$$

For example, consider the array $\{3, 2, 1; 1, 1, 3\}$ which, as we have already noted, satisfies the first three conditions. The eigenvalues of $\mathbf{B}$ are $3, -1$ and the roots of the quadratic equation $\lambda^2 + \lambda - 3 = 0$. If θ is one of the quadratic eigenvalues the corresponding eigenvector is $[1, \theta, -\theta, -1]^t$, and the 'multiplicity' is

$$12 \Big/ \left(\frac{1}{1} + \frac{\theta^2}{3} + \frac{\theta^2}{6} + \frac{1}{2} \right) = 24/(3 + \theta^2) = 24/(6 - \theta),$$

which is clearly not an integer. Thus there is no graph with the given array.

For a positive example, consider the array $\{2r, r - 1; 1, 4\}$ $(r \ge 2)$, for which the corresponding $\mathbf{B}$ matrix is

$$\begin{bmatrix} 0 & 1 & 0 \\ 2r & r & 4 \\ 0 & r-1 & 2r-4 \end{bmatrix}.$$

It is easy to verify that $k = 2r$, $k_2 = \frac{1}{2}r(r-1)$, $n = \frac{1}{2}(r+1)(r+2)$, so that conditions (1), (2) and (3) of Definition 21.5 are fulfilled.

The eigenvalues of $\mathbf{B}$ are $\lambda_0 = 2r$, $\lambda_1 = r - 2$, $\lambda_2 = -2$, and the calculation of the multiplicities goes as follows:

$$\mathbf{v}_0 = \begin{bmatrix} 1 \\ 2r \\ \frac{1}{2}r(r-1) \end{bmatrix}, \quad \mathbf{v}_1 = \begin{bmatrix} 1 \\ r-2 \\ 1-r \end{bmatrix}, \quad \mathbf{v}_2 = \begin{bmatrix} 1 \\ -2 \\ 1 \end{bmatrix}.$$

$$m(\lambda_1) = \frac{n}{(\mathbf{u}_1, \mathbf{v}_1)} = \frac{\frac{1}{2}(r+1)(r+2)}{1 + (r-2)^2/2r + (1-r)^2/\frac{1}{2}r(r-1)} = r + 1;$$

$$m(\lambda_2) = \frac{n}{(\mathbf{u}_2, \mathbf{v}_2)} = \frac{\frac{1}{2}(r+1)(r+2)}{1 + 4/2r + 1/\frac{1}{2}r(r-1)} = \frac{1}{2}(r-1)(r-2).$$

Since these values are integers, condition (4) is satisfied and the array is feasible. In fact the array is realized by the triangle graph Δ_{r+2}, as we noted in Chapter 20. (The eigenvalues and multiplicities of this graph were found in a different way in Chapter 3.)

Another example is the graph Σ representing the 27 lines on a cubic surface (Chapter 8, p. 57). This is a distance-regular graph with diameter 2 and intersection array $\{16, 5; 1, 8\}$, from which we may calculate the spectrum:

$$\text{Spec } \Sigma = \begin{pmatrix} 16 & 4 & -2 \\ 1 & 6 & 20 \end{pmatrix}.$$

These examples have diameter 2, and so they are strongly regular graphs. In that case the multiplicities can also be obtained by more elementary methods (see **3d**). But for a general distance-regular graph the multiplicity formula is invaluable.

Additional Results

21a *The spectra of Q_k and the Hamming graphs*   The eigenvalues of the k-cube Q_k are $\lambda_i = k - 2i$ ($0 \leq i \leq k$), with multiplicities $m(\lambda_i) = \binom{k}{i}$.

The k-cube is the case $q = 2$ of the *Hamming graph* $H(d, q)$, whose vertices are the q^d d-vectors with elements in a set of size q, two being adjacent when they differ in just one coordinate. The graph $H(d, q)$ is distance-transitive, with intersection array

$$\{d(q-1), (d-1)(q-1), \ldots, (q-1); 1, 2, \ldots, d\}.$$

The eigenvalues are $d(q-1) - qi$, $i = 0, 1, \ldots, d$ with multiplicities $\binom{d}{i}(q-1)^i$. The intersection array determines the Hamming graph $H(d, q)$ uniquely, except when $q = 4$; in that case there are other graphs with the same intersection array [BCN, p. 262].

21b *The spectrum of O_k*   The eigenvalues of the odd graph O_k are $\lambda_i = (-1)^i(k-i)$ $(0 \le i \le k-1)$, and

$$m(\lambda_i) = \binom{2k-1}{i} - \binom{2k-1}{i-1}.$$

21c *Elementary conditions on the intersection array*   The following conditions must be satisfied by the intersection array of any distance-regular graph. Proofs may be found in Biggs (1976).
(1)   If $a_1 = 0$ and $a_2 \ne 0$ then $a_2 \ge c_2$.
(2)   If $a_1 = 1$ then $a_2 \ge c_2$.
(3)   If $a_2 = 2$ and k is not a multiple of 3, then $c_2 \ge 2$.

21d *Integrality of all intersection numbers*   Since the matrices $\mathbf{B}_i$ are the images of the $\mathbf{A}_i$ under a faithful representation it follows that they satisfy the relation $\mathbf{B}_i = v_i(\mathbf{B})$ $(0 \le i \le d)$. Since $(\mathbf{B}_h)_{ij}$ is the number s_{hij}, it follows that each of the matrices computed by means of this formula must have integral entries.

21e *The Krein conditions*   Define

$$\mathbf{E}_i = \frac{m(\lambda_i)}{n}\mathbf{L}_i \quad (0 \le i \le d),$$

where the $\mathbf{L}_i$ are as in the proof of Theorem 21.4. The $\mathbf{E}_i$ are mutually orthogonal, idempotent, and form a basis for the adjacency algebra. This algebra is closed under the pointwise product $\circ$ of matrices, because $\mathbf{A}_i \circ \mathbf{A}_j = \delta_{ij}\mathbf{A}_j$. It follows that there are real numbers q_{hij} such that

$$\mathbf{E}_i \circ \mathbf{E}_j = \sum_h q_{hij}\mathbf{E}_h.$$

Scott (1973) observed that these *Krein parameters* must be non-negative. Thus we have a new set of 'feasibility' conditions, which can be stated explicitly as follows:

$$q_{hij} = \sum_{r=0}^{d} \frac{v_r(\lambda_h)v_r(\lambda_i)v_r(\lambda_j)}{k_r^2} \ge 0.$$

21f *An array which is not realisable*   The array $\{9, 8; 1, 4\}$ is feasible in the sense of Definition 21.5. We have

$$v_0(\lambda) = 1, \quad v_1(\lambda) = \lambda, \quad v_2(\lambda) = \frac{1}{4}(\lambda^2 - 2),$$

and the 'eigenvalues' are $9, 1, -5$ with 'multiplicities' $1, 21, 6$ respectively. The conditions given in **21c** are satisfied, and also **21d** since

$$\mathbf{B}_1 = \mathbf{B} = \begin{bmatrix} 0 & 1 & 0 \\ 9 & 0 & 4 \\ 0 & 8 & 5 \end{bmatrix}, \quad \mathbf{B}_2 = v_2(\mathbf{B}) = \begin{bmatrix} 0 & 0 & 1 \\ 0 & 8 & 5 \\ 18 & 10 & 12 \end{bmatrix}.$$

However the Krein condition $q_{222} \geq 0$, in the notation of **21e**, does not hold. An elementary proof that this array is not realisable was given by Biggs (1970).

21g *Feasibility conditions for strongly regular graphs* A strongly regular graph, as defined in **3c**, is a distance-regular graph with intersection array $\{k, k - a - 1; 1, c\}$. The eigenvalues and their multiplicities can be computed by the elementary methods described in **3d**, or by the general methods described in this chapter. A good survey is given by Seidel (1979). In addition to the 'feasibility' conditions which hold for distance-regular graphs in general, there is a useful 'absolute bound'

$$n \leq \frac{1}{2}m(m + 3),$$

where n is the number of vertices and m is the multiplicity of either one of the eigenvalues $\lambda \neq k$. For example, this test shows that the array considered in **21f** is not realisable.

21h *The friendship theorem* If, in a finite set of people, each pair of people has precisely one common friend, then someone is everyone's friend. (Friendship is interpreted as a symmetric, irreflexive relation.) The result may be proved as follows. Let Γ denote the graph whose vertices represent people and whose edges join friends. Then Γ is either a graph consisting of a number of triangles all with a common vertex or a strongly regular graph with intersection array $\{k, k - 2; 1, 1\}$. The array is not feasible, so the first possibility must hold. This is an unpublished proof of G. Higman; for other proofs see Hammersley (1981).

21i *Distance-regular and distance-transitive graphs with degree* 3 Biggs and Smith (1971) proved that there are exactly 12 distance-transitive graphs with degree 3. They are: (i) the symmetric cubic graphs with $n \leq 30$ vertices listed in **18h**, with the exception of $P(8, 3)$ and $P(12, 5)$; (ii) the threefold covering of $S(3)$ with $n = 90$ vertices described in **19c**; (iii) the expansion of **H** with $n = 102$ vertices described in **18g**.

Biggs, Boshier and Shawe-Taylor (1986) showed that in the distance-regular case there is just one other graph, which has 126 vertices (see **23b**).

21j *Perfect codes in distance-regular graphs* The definition of a perfect e-code in a graph was given in **3k**. Let $v_i(\lambda)$ be the polynomials associated with a distance-regular graph Γ, and let

$$x_i(\lambda) = v_0(\lambda) + v_1(\lambda) + \ldots + v_i(\lambda) \quad (0 \leq i \leq d).$$

If there is a perfect e-code in Γ then $x_e(\lambda)$ is a factor of $x_d(\lambda)$ in the ring of polynomials with rational coefficients. This implies that the zeros of $x_e(\lambda)$ must be eigenvalues of Γ. This result was first established by S.P. Lloyd in the 'classical' case of a cube or Hamming graph. Biggs (1973c) gave a proof for the general distance-transitive case and Delsarte (1973) proved similar results in a more general context.

21k *Sporadic groups and graphs*   Several of the sporadic simple groups can be represented as the automorphism group of a distance-transitive graph. A typical example is the distance-transitive graph with 266 vertices which has degree 11, diameter 4, and intersection array $\{11, 10, 6, 1;$ $1, 1, 5, 11\}$. The automorphism group of this graph is Janko's simple group of order 175 560. As usual, the reader should consult [BCN] for a full account.

21l *The permutation character*   If Γ is a distance-transitive graph with diameter d, then the permutation character χ corresponding to the representation of Aut(Γ) on $V\Gamma$ is the sum of $d+1$ irreducible characters:

$$\chi = 1 + \chi_1 + \ldots + \chi_d$$

and the labelling can be chosen so that the degree of χ_i is $m(\lambda_i)$ $(0 \leq i \leq d)$. This can be deduced from the results of Wielandt (1964); see also [BCN, p. 137].

22

Imprimitivity

In this chapter we investigate the relationship between primitivity and distance-transitivity. We shall prove that the automorphism group of a distance-transitive graph can act imprimitively in only two ways, both of which have simple characterizations in terms of the structure of the graph.

We begin by summarizing some terminology. If G is a group of permutations of a set X, a *block* B is a subset of X such that B and $g(B)$ are either disjoint or identical, for each g in G. If G is transitive on X, then we say that the permutation group (X, G) is *primitive* if the only blocks are the *trivial* blocks, that is, those with cardinality 0, 1 or $|X|$. If B is a non-trivial block and G is transitive on X, then each $g(B)$ is a block, and the distinct blocks $g(B)$ form a partition of X which we refer to as a *block system*. Further, G acts transitively on these blocks.

A graph Γ is said to be primitive or imprimitive according as the group $G = \mathrm{Aut}(\Gamma)$ acting on $V\Gamma$ has the corresponding property. For example, the ladder graph L_3 is imprimitive: there is a block system with two blocks, the vertices of the triangles in L_3.

Proposition 22.1 *Let Γ be a connected graph for which the group of automorphisms acts imprimitively and symmetrically (in the sense of Definition 15.5). Then a block system for the action of $\mathrm{Aut}(\Gamma)$ on $V\Gamma$ must be a colour-partition of Γ.*

Proof Suppose that $V\Gamma$ is partitioned by the block system
$$B^{(1)}, B^{(2)}, \ldots, B^{(l)}.$$

Then we may select one block, call it C, and elements $g^{(i)}$ in $\mathrm{Aut}(\Gamma)$, such that

$$B^{(i)} = g^{(i)}C \ (1 \le i \le l).$$

Suppose C contains two adjacent vertices u and v. Since Γ is symmetric, for each vertex w adjacent to u there is an automorphism g such that $g(u) = u$ and $g(v) = w$. Then u belongs to $C \cap g(C)$, and C is a block, so $C = g(C)$ and w belongs to C. Since w was any vertex adjacent to v, the set $\Gamma_1(u)$ is contained in C, and by repeating the argument we can show that $\Gamma_2(u), \Gamma_3(u), \dots$ are contained in C. Since Γ is connected, we have $C = V\Gamma$. This contradicts the hypothesis of imprimitivity, and so our assumption that C contains a pair of adjacent vertices is false. Thus C is a colour-class and since each block $B^{(i)}$ is the image of C under an automorphism, the block system is a colour-partition. □

This result is false if we assume only that the graph is vertex-transitive, rather than symmetric. The ladder graph L_3 mentioned above provides a counter-example.

The rest of this chapter is devoted to an investigation of the relationship between primitivity and distance-transitivity. We shall show that, in an imprimitive distance-transitive graph, the vertex-colouring induced by a block system is either a 2-colouring or a colouring of another quite specific kind.

Lemma 22.2 *Let Γ be a distance-transitive graph with diameter d, and suppose B is a block for the action of $\mathrm{Aut}(\Gamma)$ on $V\Gamma$. If B contains two vertices u and v such that $\partial(u,v) = j \ (1 \le j \le d)$, then B contains all the sets $\Gamma_{rj}(u)$, where r is an integer satisfying $0 \le rj \le d$.*

Proof Let w be any vertex in $\Gamma_j(u)$. Since Γ is distance-transitive there is an automorphism g such that $g(u) = u$ and $g(v) = w$. Thus u is in $B \cap g(B)$, and since B is a block, $B = g(B)$ and w is in B. So $\Gamma_j(u) \subseteq B$.

If z is in $\Gamma_{2j}(u)$, there is a vertex $y \in \Gamma_j(u)$ for which $\partial(y,z) = j$. Since $\partial(z,y) = \partial(u,y)$, and both u and y are in B, it follows by a repetition of the argument in the previous paragraph that z is in B, and so $\Gamma_{2j}(u) \subseteq B$. Further repetitions of the argument show that $\Gamma_{rj}(u) \subseteq B$ for each r such that $rj \le d$. □

For the rest of this chapter we use the symbol d' to denote the largest even integer not exceeding d.

Proposition 22.3 *Let Γ be a distance-transitive graph with diameter d and degree $k \geq 3$. Then a non-trivial block for the action of $\mathrm{Aut}(\Gamma)$ on $V\Gamma$ which contains the vertex u must be one of the following sets:*

$$B_a(u) = \{u\} \cup \Gamma_d(u), \quad B_b(u) = \{u\} \cup \Gamma_2(u) \cup \Gamma_4(u) \cup \ldots \cup \Gamma_{d'}(u).$$

Proof Suppose B is a non-trivial block containing u, and is not the set $B_a(u)$. Then B contains a vertex $v \neq u$ such that $\partial(u, v) = j < d$, and consequently $\Gamma_j(u) \subseteq B$.

Consider the numbers c_j, a_j, b_j in the intersection array of Γ. We must have $a_j = 0$, because if a_j were non-zero then B would contain two adjacent vertices, which is impossible by Proposition 22.1. Since

$$c_j + a_j + b_j = k \geq 3,$$

one of c_j, b_j is at least 2. From parts (2) and (3) of Proposition 20.4 it follows that one of c_{j+1}, b_{j-1} is at least 2, and consequently $\Gamma_j(u)$ contains a pair of vertices at distance 2. Thus B contains the set $B_b(u)$. If it contained any other vertices, it would contain two adjacent vertices and would be the trivial block $V\Gamma$. We deduce that $B = B_b(u)$, as required. $\qquad\square$

The cube Q_3 is an example of an imprimitive distance-transitive graph with diameter $d = 3$, so $d' = 2$ here. One block system consists of four sets of the form $\{u\} \cup \Gamma_3(u)$ of size two, while another block system consists of two sets of the form $\{u\} \cup \Gamma_2(u)$ of size four. This example illustrates the fact that both types of imprimitivity allowed by Proposition 22.3 can occur in the same graph.

Another instructive example is the cocktail-party graph $CP(s)$, as defined on p. 17. Here there are s blocks $u \cup \Gamma_2(u)$, each of size two, and since $d' = d = 2$ these blocks are simultaneously of type $B_a(u)$ and $B_b(u)$. The next lemma clears up this case.

Lemma 22.4 *Let Γ be a distance-transitive graph with girth 3 and diameter $d \geq 2$, in which the set*

$$B_b(u) = \{u\} \cup \Gamma_2(u) \cup \ldots \cup \Gamma_{d'}(u)$$

is a block. Then $d = 2$ and consequently $B_b(u) = \{u\} \cup \Gamma_2(u) = B_a(u)$.

Proof Since Γ contains triangles and is distance-transitive, every ordered pair of adjacent vertices belongs to a triangle. Choose adjacent vertices $v_1 \in \Gamma_1(u)$, $v_2 \in \Gamma_2(u)$; then there is some vertex z such that $v_1 v_2 z$ is a triangle. If z were in $\Gamma_2(u)$, then $B_b(u)$ would contain adjacent vertices, contrary to Proposition 22.1. Thus z must be in $\Gamma_1(u)$.

If $d \geq 3$, we can find a vertex $v_3 \in \Gamma_3(u)$ which is adjacent to v_2

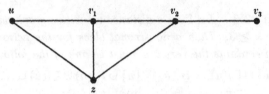

Figure 18: illustrating the proof of Lemma 22.4.

(Figure 18). But then $\Gamma_2(v_3)$ contains the adjacent vertices v_1 and z, and if h is an automorphism of Γ taking u to v_3, $h(B_b(u))$ is a block containing adjacent vertices, again contradicting Proposition 22.1. Thus we must have $d = 2$.   □

Proposition 22.5   *Let Γ be a distance-transitive graph with diameter $d \geq 3$ and degree $k \geq 3$. Then*

$$X = B_b(u) = \{u\} \cup \Gamma_2(u) \cup \ldots \cup \Gamma_{d'}(u)$$

is a block if and only if Γ is bipartite.

Proof   Suppose Γ is bipartite. If X is not a block, then there is an automorphism g of Γ such that X and $g(X)$ intersect but are not identical. This would imply that there are vertices x and y in X, for which $g(x) \in X$ but $g(y) \notin X$, so that $\partial(x, y)$ is even and $\partial(g(x), g(y))$ is odd. From this contradiction we conclude that X is a block.

Conversely, suppose X is a block. A minimal odd cycle in Γ has length $2j + 1$ greater than 3, by Lemma 22.4. We may suppose this cycle to be $uu_1 \ldots w_1 v_1 v_2 w_2 \ldots u_2 u$, where

$$u_1, u_2 \in \Gamma_1(u), \quad w_1, w_2 \in \Gamma_{j-1}(u), \quad v_1, v_2 \in \Gamma_j(u),$$

and if $j = 2$, then $u_1 = w_1$ and $u_2 = w_2$. If j is even, then X contains the adjacent vertices v_1 and v_2, and so $X = V\Gamma$, a contradiction. If j is odd we have, for $i = 1, 2$, $\partial(u, w_i) = \partial(u_i, v_i)$, and so there is an automorphism h_i taking u to u_i and w_i to v_i. Thus $Y_i = h_i(X)$ is a block containing u_i and v_i. But, since Γ contains no triangles, $\partial(u_1, u_2) = 2$ and so $u_2 \in Y_1$. Consequently $Y_1 = Y_2$ and we have adjacent vertices v_1, v_2 in Y_1, so that $Y_1 = V\Gamma, X = V\Gamma$. From this contradiction it follows that Γ has no odd cycles and is bipartite.   □

Lemma 22.4 and Proposition 22.5 lead to the conclusion that, if a block of the type $B_b(u)$ exists in a distance-transitive graph Γ, then either $d = 2$, in which case the block is also of type $B_a(u)$, or $d \geq 3$ and Γ is bipartite. The complete tripartite graphs $K_{r,r,r}$ are examples of the first case and are clearly not bipartite.

We shall now show that graphs which have blocks of type $B_a(u)$ can also be given a simple graph-theoretical characterization.

Definition 22.6 A graph of diameter d is said to be *antipodal* if, for any vertices u, v, w such that $\partial(u, v) = \partial(u, w) = d$, it follows that $\partial(v, w) = d$ or $v = w$.

The cubes Q_k are trivially antipodal, since every vertex has a unique vertex at maximum distance from it; these graphs are at the same time bipartite. The dodecahedron is also trivially antipodal, but it is not bipartite. Examples of graphs which are non-trivially antipodal and not bipartite are the complete tripartite graphs $K_{r,r,r}$, which have diameter 2, and the line graph of Petersen's graph, which has diameter 3.

Proposition 22.7 *A distance-transitive graph Γ of diameter d has a block $B_a(u) = \{u\} \cup \Gamma_d(u)$ if and only if Γ is antipodal.*

Proof Suppose Γ is antipodal. Then if x is in $B_a(u)$, it follows that $B_a(u) = \{x\} \cup \Gamma_d(x) = B_a(x)$. Consequently, if g is any automorphism of Γ, and z is in $B_a(u) \cap g(B_a(u))$ then

$$B_a(u) = \{z\} \cup \Gamma_d(z) = g(B_a(u)),$$

so that $B_a(u)$ is a block.

Conversely, suppose $B_a(u)$ is a block, and v, w belong to $\Gamma_d(u)$ ($v \neq w$). Let $\partial(v, w) = j$ ($1 \leq j \leq d$), and let h be any automorphism of Γ such that $h(v) = u$. Then $h(w)$ is in $\Gamma_j(u)$. Also $h(w)$ belongs to $h(B_a(u)) = B_a(u)$, since $h(B_a(u))$ intersects $B_a(u)$ (u is in both sets) and $B_a(u)$ is a block. This is impossible for $1 \leq j < d$, so that $\partial(v, w) = d$, and Γ is antipodal.

Theorem 22.8 (Smith 1971) *An imprimitive distance-transitive graph with degree $k \geq 3$ is either bipartite or antipodal. (Both possibilities can occur in the same graph.)*

Proof A non-trivial block is either of the type $B_a(u)$ or $B_b(u)$. In the case of a block of type $B_b(u)$, Proposition 22.5 tells us that either the graph is bipartite, or its diameter is less than 3. If the diameter is 1, then the graph is complete, and consequently primitive. If the diameter is 2, a block of type $B_b(u)$ is also of type $B_a(u)$. Consequently, if the graph is not bipartite, it must be antipodal. $\square$

The notion of primitivity can be defined without reference to a group action, in the following way. Given a graph Γ with diameter d, let Γ_i ($1 \leq i \leq d$) be the graph whose vertices are the same as those of Γ, two vertices being adjacent in Γ_i if and only if they are at distance i in Γ.

Then Γ is said to be *imprimitive* if any of the graphs Γ_i is disconnected. It is easy to see that for a bipartite graph Γ_2 has two components, and for an antipodal graph Γ_d is the disjoint union of complete graphs. Using this definition Smith's theorem and its proof can be extended to distance-regular graphs (see [BCN, p. 140]).

The complete graphs are primitive and distance-transitive. Other families with the same properties are line graphs of a certain kind. Apart from these families, primitive distance-transitive graphs are scarce, and we give them a special name.

Definition 22.9 An *automorphic* graph is a distance-transitive graph which is primitive and not a complete graph or a line graph.

For instance, of the 12 distance-transitive graphs with degree 3 (**21i**), only three are automorphic. They are Petersen's graph, Coxeter's graph (the expansion of **Y** with 28 vertices), and the expansion of **H** with 102 vertices. The odd graph O_4 is the only automorphic graph with degree 4. Many more details may be found in [BCN].

Additional Results

22a *The derived graph of an antipodal graph*   Let Γ be a distance-transitive antipodal graph, with degree k and diameter $d > 2$. Define the *derived graph* Γ' by taking the vertices of Γ' to be the blocks $\{u\} \cup \Gamma_d(u)$ in Γ, two blocks being joined in Γ' whenever they contain adjacent vertices of Γ. Then Γ' is a distance-transitive graph with degree k and diameter equal to $\lfloor d/2 \rfloor$ (Smith 1971).

22b *The icosahedron and the dodecahedron*   The icosahedron I and the dodecahedron D are distance-transitive with

$$\iota(I) = \{5, 2, 1; 1, 2, 5\}; \quad \iota(D) = \{3, 2, 1, 1, 1; 1, 1, 1, 2, 3\}.$$

Both graphs are antipodal, and the derived graphs are K_6 and O_3.

22c *The intersection array of an antipodal covering*   We can look at the construction in **22a** from the opposite point of view, as follows. A distance-regular graph $\widetilde{\Gamma}$ is an *antipodal r-fold covering* of the distance-regular graph Γ if $\widetilde{\Gamma}$ is antipodal, its derived graph is Γ, and $|V\widetilde{\Gamma}| = r|V\Gamma|$. It turns out that the intersection array of $\widetilde{\Gamma}$ is related to the intersection array $\{k, b_1, \ldots, b_{d-1}; 1, c_2, \ldots, c_d\}$ of Γ in one of two ways. Either (i) $\widetilde{\Gamma}$ has even diameter $2d > 2$ and

$$\iota(\widetilde{\Gamma}) = \{k, b_1, \ldots, b_{d-1}, (r-1)c_d/r, c_{d-1}, \ldots, c_2, 1;$$

$$1, c_2, \ldots, c_{d-1}, c_d/r, b_{d-1}, \ldots, b_1, k\},$$

or (ii) $\widetilde{\Gamma}$ has odd diameter $2d+1$ and, for some positive integer t such that $(r-1)t \leq \min(b_{d-1}, a_d)$ and $c_d \leq t$, we have

$$\iota(\widetilde{\Gamma}) = \{k, b_1, \ldots, b_{d-1}, (r-1)t, c_d, c_{d-1}, \ldots, c_2, 1;$$
$$1, c_2, \ldots, c_d, t, b_{d-1}, \ldots, b_1, k\}.$$

Clearly, the total number of possibilities is finite, and $r \leq k$ in any case.

22d *Antipodal coverings of $K_{k,k}$* Let Γ be a distance-regular graph which is an antipodal r-fold covering of $K_{k,k}$. Then it follows from **22c** that r must divide k, and if $rt = k$ the intersection array for Γ is

$$\{k, k-1, k-t, 1; 1, t, k-1, k\}.$$

This array is feasible (provided that r divides k) and the spectrum of Γ is

$$\text{Spec } \Gamma = \begin{pmatrix} k & \sqrt{k} & 0 & -\sqrt{k} & -k \\ 1 & k(r-1) & 2(k-1) & k(r-1) & 1 \end{pmatrix}.$$

In the case $r = k$, the existence of Γ implies the existence of a projective plane of order k (Gardiner 1974).

22e *Distance-regular graphs with diameter three* A distance-regular graph with diameter three is antipodal, bipartite, or primitive (in the extended sense defined on p. 177). In the antipodal case the intersection array is of the form $\{k, (r-1)\gamma, 1; 1, \gamma, k\}$, and the graph is an antipodal r-fold covering of K_{k+1}. This case has been the subject of several papers: see Biggs (1982c), Cameron (1991), Godsil and Hensel (1992). In the bipartite case the intersection array is of the form $\{k, k-1, k-\lambda; 1, \lambda, k\}$, and the graph is the incidence graph of a symmetric 2-design with parameters (v, k, λ), where $v = k(k-1)/\lambda + 1$. Several families of primitive graphs are known, and some sporadic ones [BCN, pp. 425–431].

22f *An automorphic graph with $k = 5$ and $d = 3$* Let $L = \{a, b, c, d, e, f\}$, and $N = \{1, 2, 3, 4, 5, 6\}$. The following table establishes a bijection between the 15 single-transpositions on L, and the 15 triple-transpositions on N.

$$\begin{array}{lll}
(ab) \mapsto (15)(23)(46) & (ac) \mapsto (14)(26)(35) & (ad) \mapsto (13)(24)(56) \\
(ae) \mapsto (12)(36)(45) & (af) \mapsto (16)(25)(34) & (bc) \mapsto (12)(34)(56) \\
(bd) \mapsto (14)(25)(36) & (be) \mapsto (16)(24)(35) & (bf) \mapsto (13)(26)(45) \\
(cd) \mapsto (16)(23)(45) & (ce) \mapsto (13)(25)(46) & (cf) \mapsto (15)(24)(36) \\
(de) \mapsto (15)(26)(34) & (df) \mapsto (12)(35)(46) & (ef) \mapsto (14)(23)(56)
\end{array}$$

Define a graph Γ whose vertex-set is $L \times N$, and in which (l_1, n_1) is adjacent to (l_2, n_2) if and only if the transposition (n_1, n_2) is one of those corresponding to (l_1, l_2). Then Γ is an automorphic graph with degree 5 and diameter 3. Its intersection array is $\{5, 4, 2; 1, 1, 4\}$ and its automorphism group is Aut S_6.

23

Minimal regular graphs with given girth

Results on the feasibility of intersection arrays can be applied to a wide range of combinatorial problems. The last chapter of this book deals with a graph-theoretical problem which has been the subject of much research. We shall study regular graphs whose degree ($k \geq 3$) and girth ($g \geq 3$) are given. For all such values of k and g there is at least one graph with these properties (Sachs 1963), and so it makes sense to ask for the smallest one. We note that when $k = 2$ the cycle graphs provide the complete answer to the problem, and so we shall be concerned primarily with the case $k \geq 3$.

Proposition 23.1 (1) *The number of vertices in a graph with degree k and odd girth $g = 2d + 1$ is at least*

$$n_0(k, g) = 1 + k + k(k - 1) + \ldots + k(k - 1)^{\frac{1}{2}(g-3)}.$$

If there is such a graph, having exactly $n_0(k, g)$ vertices, then it is distance-regular with diameter d, and its intersection array is

$$\{k, k - 1, k - 1, \ldots, k - 1; 1, 1, 1, \ldots, 1\}.$$

(2) *The number of vertices in a graph with degree k and even girth $g = 2d$ is at least*

$$n_0(k, g) = 1 + k + k(k - 1) + \ldots + k(k - 1)^{\frac{1}{2}g-2} + (k - 1)^{\frac{1}{2}g-1}.$$

If there is such a graph, having exactly $n_0(k, g)$ vertices, then it is bipartite and distance-regular with diameter d; its intersection array is

$$\{k, k - 1, k - 1, \ldots, k - 1; 1, 1, 1, \ldots, 1, k\}.$$

Proof (1) Suppose that Γ is a graph with degree k and girth $g = 2d+1$, and let (u,v) be any pair of vertices such that $\partial(u,v) = j$ $(1 \le j \le d)$. The number of vertices in $\Gamma_{j-1}(v)$ adjacent to u is 1, otherwise we should have a cycle of length at most $2j < 2d + 1$ in Γ. Using the standard notation (Definition 20.5) we have shown the existence of the numbers $c_1 = 1, \ldots, c_d = 1$. Similarly, if $1 \le j \le d$, then there are no vertices in $\Gamma_j(v)$ adjacent to u, otherwise we should have a cycle of length at most $2j+1 < 2d+1$. This means that $a_j = 0$ and consequently $b_j = k - a_j - c_j = k - 1$, for $1 \le j < d$. It follows that the diameter of Γ is at least d, and that Γ has at least $n_0(k,g)$ vertices. If Γ has just $n_0(k,g)$ vertices, its diameter must be precisely d, which implies that $a_d = 0$, and Γ has the stated intersection array.

(2) In this case the argument proceeds as in (1), except that c_d may be greater than one. Now the recurrence for the numbers $k_i = |\Gamma_i(v)|$ shows that k_d is smallest when $c_d = k$; if this is so, then Γ has at least $n_0(k,g)$ vertices. If Γ has exactly $n_0(k,g)$ vertices, then its diameter is d, and it has the stated intersection array. The form of this array shows that Γ has no odd cycles, and so it is bipartite. $\square$

Definition 23.2 A graph with degree k, girth g, and such that there are no smaller graphs with the same degree and girth, is called a (k,g)-*cage*. A (k,g)-cage with $n_0(k,g)$ vertices is said to be a *Moore graph* if g is odd, and a *generalized polygon graph* if g is even. (The reasons for the apparently bizarre terminology are historical, and may be found in the references given below.)

We have already remarked that a (k,g)-cage exists for all $k \ge 3$ and $g \ge 3$. For example, Petersen's graph O_3 is the unique $(3,5)$-cage: it has 10 vertices and $n_0(3,5) = 10$, so it is a Moore graph. On the other hand, the unique $(3,7)$-cage has 24 vertices (see **23c**) and $n_0(3,7) = 22$, so there is no Moore graph in this case. The main result of this chapter is that Moore graphs and generalized polygon graphs are very rare.

In the cases $g = 3$ and $g = 4$ the intersection arrays in question are

$$\{k; 1\} \quad \text{and} \quad \{k, k-1; 1, k\}$$

and these are feasible for all $k \ge 3$. It is very easy to see that each array has a unique realisation – the complete graph K_{k+1} and the complete bipartite graph $K_{k,k}$, respectively. Thus, when $g = 3$ we have a unique Moore graph K_{k+1}, and when $g = 4$ we have a unique generalized polygon graph $K_{k,k}$.

When $g \ge 5$ the problem is much more subtle, both in the technical details and in the nature of the solution. The results are due to a number of mathematicians. The generalized polygon case was essentially solved by Feit and Higman (1964); the Moore graph case was investigated by Hoffman and Singleton (1960), Vijayan (1972), Damerell (1973) and Bannai and Ito (1973).

We shall apply the algebraic techniques developed in Chapter 21 to both cases, in a uniform manner. Specifically we investigate the feasibility of the intersection matrix

$$\mathbf{B} = \begin{bmatrix} 0 & 1 & & & & & \\ k & 0 & 1 & & & & \\ & k-1 & 0 & . & & & \\ & & k-1 & . & . & & \\ & & & . & . & 1 & \\ & & & & . & 0 & c \\ & & & & & k-1 & k-c \end{bmatrix},$$

which subsumes, by putting $c = 1$ and $c = k$, the intersection matrices of Moore graphs and generalized polygon graphs.

Suppose that λ is an eigenvalue of $\mathbf{B}$ and that the corresponding standard left eigenvector is $\mathbf{u}(\lambda) = [u_0(\lambda), u_1(\lambda), \ldots, u_d(\lambda)]$. Then, from the equations $\mathbf{u}(\lambda)\mathbf{B} = \lambda\mathbf{u}(\lambda)$ and $u_0(\lambda) = 1$, we deduce that $u_1(\lambda) = \lambda/k$ and

$$(*) \qquad (k-1)u_i(\lambda) - \lambda u_{i-1}(\lambda) + u_{i-2}(\lambda) = 0 \quad (i = 2, 3 \ldots, d),$$

$$(**) \qquad cu_{d-1}(\lambda) + (k - c - \lambda)u_d(\lambda) = 0.$$

The equations $(*)$ give a recursion which enables us to express $u_i(\lambda)$ as a polynomial of degree i in λ for $0 \leq i \leq d$. The equation $(**)$ then becomes a polynomial equation of degree $d + 1$ in λ. In fact $(**)$ represents the condition that λ is an eigenvalue; it is the characteristic equation of $\mathbf{B}$.

Put $q = \sqrt{k-1}$ and suppose that $|\lambda| < 2q$, so that we may write $\lambda = 2q\cos\alpha$ for some α, $0 < \alpha < \pi$ (this assumption will be justified in the course of the ensuing argument). The solution to the recursion $(*)$ can be found explicity:

$$u_i = \frac{q^2 \sin(i+1)\alpha - \sin(i-1)\alpha}{kq^i \sin\alpha} \quad (1 \leq i \leq d).$$

Lemma 23.3 *With the above notation, the number $2q\cos\alpha$ is an eigenvalue of $\mathbf{B}$ if and only if*

$$q\sin(d+1)\alpha + c\sin d\alpha + \left(\frac{c-1}{q}\right)\sin(d-1)\alpha = 0.$$

Proof The stated equation results from substituting the explicit forms of u_{d-1} and u_d in the equation $(**)$, which is the characteristic equation of $\mathbf{B}$. □

Proposition 23.4   (1) *Let* $g = 2d$ *and suppose* Γ *is a generalized polygon graph for the values* (k, g). *Then* Γ *has* $d+1$ *distinct eigenvalues:*
$$k, -k, \quad 2q \cos \pi j/d \quad (j = 1, 2, \ldots, d-1).$$
(2) *Let* $g = 2d + 1$ *and suppose* Γ *is a Moore graph for the values* (k, g). *Then* Γ *has* $d + 1$ *distinct eigenvalues:*
$$k, \quad 2q \cos \alpha_j \quad (j = 1, 2, \ldots, d),$$
where the numbers $\alpha_1, \ldots, \alpha_d$ *are the distinct solutions in the interval* $0 < \alpha < \pi$ *of the equation* $q \sin(d + 1)\alpha + \sin d\alpha = 0$.

Proof   (1) The existence of the eigenvalues k and $-k$ follows from the fact that Γ is k-regular and bipartite. Now the eigenvalues of Γ are (by Proposition 21.2) the $d + 1$ eigenvalues of its intersection matrix, which is the matrix given above with $c = k$. In that case, $\lambda = 2q \cos \alpha$ is an eigenvalue of $\mathbf{B}$ if and only if
$$q \sin(d + 1)\alpha + k \sin d\alpha + q \sin(d - 1)\alpha = 0.$$
This reduces to $(2q \cos \alpha + k) \sin d\alpha = 0$, and since $|k/2q| > 1$ when $k \geq 3$, the only possibility is that $\sin d\alpha = 0$. Thus in the range $0 < \alpha < \pi$ there are $d - 1$ solutions $\alpha = \pi j/d$, corresponding to $j = 1, \ldots, d - 1$, and we have the required total of $d + 1$ eigenvalues in all.

(2) Since Γ is k-regular, k is an eigenvalue. As in (1), we now seek eigenvalues $\lambda = 2q \cos \alpha$ of $\mathbf{B}$, this time with $c = 1$. The equation of Lemma 23.3 reduces to
$$\Delta \equiv q \sin(d + 1)\alpha + \sin d\alpha = 0.$$
For $1 \leq j \leq d$, Δ is strictly positive at $\theta_j = (j - \frac{1}{2})\pi/(d+1)$ and strictly negative at $\phi_j = (j + \frac{1}{2})\pi/(d+1)$. Hence there is a zero α_j of Δ in each one of the d intervals (θ_j, ϕ_j). Thus we have the required total of $d + 1$ eigenvalues in all. $\qquad\square$

We now have enough information to calculate the multiplicities of the eigenvalues and to test the feasibility of the corresponding intersection array. Suppose that λ is an eigenvalue of $\mathbf{B}$. The multiplicity of λ as an eigenvalue of the putative graph is given by Theorem 21.4: $m(\lambda) = n/(\mathbf{u}(\lambda), \mathbf{v}(\lambda))$. We shall use this in the form $m(\lambda) = n/\sum k_i u_i(\lambda)^2$. For our matrix $\mathbf{B}$ we have $k_0 = 1$, $k_i = k(k - 1)^{i-1}$ $(1 \leq i \leq d - 1)$, and $k_d = c^{-1}k(k - 1)^{d-1}$. Also, for an eigenvalue $\lambda = 2q \cos \alpha$ we have
$$k_i u_i(\lambda)^2 = kq^{2i-2} \left(\frac{q^2 \sin(i + 1)\alpha - \sin(i - 1)\alpha}{kq^i \sin \alpha} \right)^2$$
$$= (2hk \sin^2 \alpha)^{-1}(E + F \cos 2i\alpha + G \sin 2i\alpha) \quad (1 \leq i < d),$$
where we have written
$$h = q^2 = k - 1, \quad E = (h^2 + 1) - 2h \cos 2\alpha,$$

$$F = 2h - (h^2 + 1)\cos 2\alpha, \quad G = (h^2 - 1)\sin 2\alpha.$$

Allowing for the anomalous form of k_d by means of a compensating term, we can sum the trigonometric series involved in $\sum k_i u_i(\lambda)^2$ and obtain:

$$1 + (2hk\sin^2\alpha)^{-1}\bigg[dE + \{F\cos(d+1)\alpha + G\sin(d+1)\alpha\}\frac{\sin d\alpha}{\sin\alpha}$$

$$+(c^{-1} - 1)(E + F\cos 2d\alpha + G\sin 2d\alpha)\bigg].$$

Fortunately this expression can be simplified considerably in the two cases, $c = 1$ and $c = k$, which are of particular interest.

Proposition 23.5 *If $\lambda \neq \pm k$ is an eigenvalue of a generalized polygon graph with girth $g = 2d$, then its multiplicity is given by*

$$m(\lambda) = \frac{nk}{g}\left(\frac{4h - \lambda^2}{k^2 - \lambda^2}\right) \quad (h = k - 1).$$

If $\lambda \neq k$ is an eigenvalue of a Moore graph with girth $g = 2d + 1$, then its multiplicity is given by

$$m(\lambda) = \frac{nk}{g}\left(\frac{4h - \lambda^2}{(k - \lambda)(f + \lambda)}\right) \quad (h = k - 1, \ f = k + (k - 2)/g).$$

Proof In the case of even girth, $c = k$ and we know that $\lambda = 2q\cos\alpha$ is an eigenvalue if and only if $\sin d\alpha = 0$. In this case the expression for $\sum k_i u_i(\lambda)^2$ becomes

$$1 + (2hk\sin^2\alpha)^{-1}[dE + hk^{-1}(E + F)] = (2hk\sin^2\alpha)^{-1}dE.$$

On putting $2d = g$, $\lambda = 2q\cos\alpha$ this leads to the formula given.

In the case of odd girth, $c = 1$, and we know that $\lambda = 2q\cos\alpha$ is an eigenvalue if and only if

$$q\sin(d+1)\alpha + \sin d\alpha = 0.$$

From this equation we have

$$\tan d\alpha = \frac{-q\sin\alpha}{1 + q\cos\alpha}; \quad \sin d\alpha = \frac{-q\sin\alpha}{\sqrt{k + \lambda}}.$$

$$\sin(d+1)\alpha = \frac{\sin\alpha}{\sqrt{k + \lambda}}; \quad \cos(d+1)\alpha = \frac{q + \cos\alpha}{\sqrt{k + \lambda}}.$$

Substituting for the relevant quantities in the general expression, and putting $g = 2d + 1$, we obtain, after some algebraic manipulation, the stated formula. $\square$

We are now ready for the main theorem, which is the result of the combined efforts of the mathematicians mentioned earlier in this chapter.

Theorem 23.6 *The intersection array for a generalized polygon graph with $k \geq 3$, $g \geq 4$ is feasible if and only if $g \in \{4, 6, 8, 12\}$. The intersection array for a Moore graph with $k \geq 3$, $g \geq 5$ is feasible if and only if $g = 5$ and $k \in \{3, 7, 57\}$.*

Proof Suppose g is even, $g = 2d$. Then a generalized polygon graph has $d - 1$ eigenvalues $\lambda_j = 2q \cos(\pi j/d)$ with multiplicities

$$m(\lambda_j) = \frac{nk}{g} \left(\frac{4h - \lambda_j^2}{k^2 - \lambda_j^2} \right).$$

If $m(\lambda_1)$ is a positive integer, λ_1^2 is rational, which means that $\cos 2\pi/d$ is rational. But it is well known (see for example, *Irrational Numbers* by I. Niven (Wiley, 1956), p. 37) that this is so if and only if $d \in \{2, 3, 4, 6\}$.

The case when g is odd presents more problems. We shall deal with $g = 5$ and $g = 7$ separately, and then dispose of $g \geq 9$. Suppose $g = 5$. Then the characteristic equation

$$q \sin 3\alpha + \sin 2\alpha = 0$$

reduces, in terms of $\lambda = 2q \cos \alpha$, to $\lambda^2 + \lambda - (k - 1) = 0$. Thus there are two eigenvalues $\lambda_1 = \frac{1}{2}(-1 + \sqrt{D})$ and $\lambda_2 = \frac{1}{2}(-1 - \sqrt{D})$, where $D = 4k - 3$. We have $n = 1 + k^2$ and putting this in the formula for $m(\lambda)$ we get

$$m(\lambda) = \frac{(k + k^3)(4k - 4 - \lambda^2)}{(k - \lambda)(6k - 2 + 5\lambda)}.$$

If $\sqrt{D}$ is irrational, we multiply out the expression above, substitute $\lambda = \frac{1}{2}(-1 \pm \sqrt{D})$ and equate the coefficients of $\sqrt{D}$. This gives $5m + k - 2 = k + k^3$, where $m = m(\lambda_1) = m(\lambda_2)$. But there are three eigenvalues in all: k, λ_1, λ_2 with multiplicities $1, m, m$; hence $1 + 2m = n = 1 + k^2$. Thus $5k^2 - 4 = 2k^3$, which has no solution for $k \geq 3$. Consequently $\sqrt{D}$ must be rational, $s = \sqrt{D}$, say. Then $k = \frac{1}{4}(s^2 + 3)$ and substituting for λ_1 and k in terms of s in the expression for $m_1 = m(\lambda_1)$ we obtain the following polynomial equation in s:

$$s^5 + s^4 + 6s^3 - 2s^2 + (9 - m_1)s - 15 = 0.$$

It follows that s must be a divisor of 15, and the possibilities are $s = 1, 3, 5, 15$, giving $k = 1, 3, 5, 57$. The first possibility is clearly absurd, but the three others do lead to feasible intersection arrays.

Suppose $g = 7$. Then the characteristic equation

$$q \sin 4\alpha + \sin 3\alpha = 0$$

reduces, in terms of $\lambda = 2q \cos \alpha$, to $\lambda^3 + \lambda^2 - 2(k-1)\lambda - (k-1) = 0$. This equation has no rational roots (and consequently no integral roots), since we may write it in the form $k - 1 = \lambda^2(\lambda + 1)/(2\lambda + 1)$, and if any prime divisor of $2\lambda+1$ divides $x = \lambda$ or $\lambda+1$ it must divide $2\lambda+1-x = \lambda+1$ or λ, which is impossible. So the roots $\lambda_1, \lambda_2, \lambda_3$ are all irrational, and their multiplicities are all equal, to m say. Then $1+3m = n = 1+k-k^2+k^3$, whereas $k + m(\lambda_1 + \lambda_2 + \lambda_3) = \text{tr}\mathbf{A} = 0$. But $\lambda_1 + \lambda_2 + \lambda_3 = -1$, hence

$$m = k = \frac{1}{3}(k^3 - k^2 + k),$$

which is impossible for $k \geq 3$. Thus there are no Moore graphs when $g = 7$.

Suppose $g \geq 9$. We obtain a contradiction here by proving first that $-1 < \lambda_1 + \lambda_d < 0$, and then showing that all eigenvalues must in fact be integers. (The argument just fails in the case $k = 3$, $g = 9$, but this can be discarded by an explicit calculation of the 'multiplicities'.)

Let α_i $(1 \leq i \leq d)$ be the roots of

$$\Delta \equiv q \sin(d + 1)\alpha + \sin d\alpha = 0,$$

and set $\omega = \pi/(d+1)$. The proof of Proposition 23.4 showed that α_1 lies between $\omega/2$ and $3\omega/2$, and these bounds can be improved by noting that Δ is positive at ω and negative at $\omega(1 + 1/2q)$. Thus $\omega < \alpha_1 < \omega(1 + 1/2q)$ and

$$\begin{aligned}
0 < 2q \cos\omega - 2q \cos\alpha_1 &< 2q \cos\omega - 2q \cos\omega(1 + 1/2q) \\
&= 2q \cos\omega(1 - \cos\omega/2q) + 2q \sin\omega \sin(\omega/2q) \\
&< 2q \times \frac{1}{2}(\omega/2q)^2 + 2q\omega(\omega/2q) \\
&= (1/4q + 1)\omega^2 \\
&< 5\omega^2/4.
\end{aligned}$$

In a similar way it can be shown that $d\omega < \alpha_d < \omega(d + 1/2q)$, and

$$0 < 2q \cos d\omega - 2q \cos\alpha_d < \omega^2.$$

Adding the two inequalities, and noting that

$$\lambda_1 = 2q \cos\alpha_1, \quad \lambda_d = 2q \cos\alpha_d, \quad \cos d\omega = -\cos\omega,$$

we have

$$-9\omega^2/4 < \lambda_1 + \lambda_d < 0.$$

Now $\omega^2 = \pi^2/(d + 1)^2 \leq \pi^2/5^2 < 4/9$, so $-1 < \lambda_1 + \lambda_d < 0$, as promised.

To show that the eigenvalues must be integers we note first that since the characteristic equation is monic with integer coefficients, the eigenvalues are algebraic integers. The formula for $m(\lambda)$ is the quotient of two quadratic expressions in λ, and so $m(\lambda)$ is integral only if λ is, at worst, a quadratic irrational. Suppose λ is a quadratic irrational. Then

$$R(\lambda) = gm(\lambda)/nk = (4h - \lambda^2)/(k - \lambda)(f + \lambda)$$

is rational number, and this equation can be written in the form

$$(R(\lambda) - 1)\lambda^2 + R(\lambda)(f - k)\lambda - (R(\lambda)fk - 4h) = 0.$$

But this must be a multiple of the minimal equation for λ, which is monic with integer coefficients. In particular,

$$S(\lambda) = \frac{(f - k)R(\lambda)}{R(\lambda) - 1} = \frac{4h - \lambda^2}{\lambda - t} \quad \text{where} \quad t = \frac{fk - 4h}{f - k},$$

must be an integer. However, $f = k + (k - 2)/g > k$, so $t > (k^2 - 4h)/(f - k) = g(k - 2)$, and consequently $|\lambda - t| > g(k - 2) - k$, since $|\lambda| < k$. Thus

$$|S(\lambda)| \le \frac{4h}{|\lambda - t|} \le \frac{4k - 4}{g(k - 2) - k} < 1,$$

for all $k \ge 3, g \ge 9$ (except when $k = 3$, $g = 9$, as we have already noted). Since $S(\lambda)$ is to be an integer, we must have $S(\lambda) = 0$, which leads to the absurdity $R(\lambda) = m(\lambda) = 0$. Thus all eigenvalues λ must be integers, which is incompatible with the inequality $-1 < \lambda_1 + \lambda_d < 0$, and consequently disposes of all cases with $g \ge 9$. $\qquad\square$

The question of the existence of graphs allowed by Theorem 23.6 is a difficult one, and it contains some celebrated unsolved problems. In the case of even girth $g = 2d$, we can relate the problem to existence of a structure known as a *generalized d-gon*, defined as follows.

Let (P, L, I) be an incidence system consisting of two disjoint finite sets, P (points) and L (lines), and an incidence relation I between points and lines. A sequence whose terms are alternately points and lines, each term being incident with its successor, is called a chain; it is a proper chain if there are no repeated terms, except possibly when the first and last terms are identical (when we speak of a closed chain). A (non-degenerate) generalized d-gon is an incidence system with the properties: (a) each pair of elements of $P \cup L$ is joined by a chain of length at most d; (b) there is a pair of elements of $P \cup L$ for which there is no proper chain of length less than d joining them; (c) there are no closed chains of length less than $2d$.

Denote by $G_d(s, t)$ a generalized d-gon with s points on each line and t lines through each point. Given a $G_d(k, k)$, the graph whose vertex-set is $P \cup L$ and whose edge-set consists of incident pairs is a $(k, 2d)$-cage with $n_0(k, 2d)$ vertices. The converse is also true. Thus our 'generalized polygon graphs' are just the incidence graphs of generalized d-gons with $s = t$.

It is easy to construct a $G_2(k,k)$ for all $k \geq 2$; the corresponding graph is the complete bipartite graph $K_{k,k}$. A $G_3(k,k)$ is simply a projective plane with k points on each line. So the existence problem for generalized polygon graphs of girth 6 is covered by the known results on projective planes, a fact noted by Singleton (1966). There is at least one such plane whenever $k-1$ is a prime power, and none are known for which $k-1$ is not a prime power. Generalized quadrangles $G_4(k,k)$ are also known to exist for all prime power values of $k-1$, and generalized hexagons $G_6(k,k)$ exist whenever $k-1$ is an odd power of 3. Benson (1966) was the first to construct the graphs corresponding to the the last two cases.

In the case of odd girth $g > 3$, the only Moore graphs allowed by Theorem 23.6 are those with $g = 5$ and $k \in \{3,7,57\}$. The graph with $k = 3$ is Petersen's graph. The graph with $k = 7$ was constructed and proved unique by Hoffman and Singleton (1960); a construction is given in **23d**. The existence of a graph with $k = 57$ remains an enigma: the results of Aschbacher (1971) show that such a graph cannot be distance-transitive, and so the construction, if there is one, is certain to be very complicated.

Additional Results

23a *Moore graphs and generalized polygon graphs with degree* 3   In the case $k = 3$, the Moore graphs of girth 3 and girth 5 (K_4 and O_3) exist and are unique. There are no other Moore graphs of degree 3, by Theorem 23.6. The generalized polygon graphs of girth 4, 6, 8 and 12 exist and are unique. They are $K_{3,3}$, Heawood's graph $S(7)$, Tutte's graph Ω, and the incidence graph of the unique generalized hexagon with 63 points and 63 lines (see **23b**).

23b *The* $(3,12)$*-cage*   A direct construction of the generalized hexagon graph of degree 3 is as follows. Given a unitary polarity of the projective plane $PG(2,3^2)$, there are 63 points of the plane which do not lie on their polar lines, and they form 63 self-polar triangles (Edge 1963). The $(3,12)$-cage is the graph whose 126 vertices are these 63 points and 63 triangles, with adjacent vertices corresponding to an incident (point, triangle) pair.

This graph is not vertex-transitive, since there is no automorphism taking a 'point' vertex to a 'triangle' vertex. However, it follows from Proposition 23.1 that it is distance-regular.

23c *Cages with degree 3 and $g \leq 12$*   All cases except $g = 7, 9, 10$ and 11 have been covered above. In these cases we know from the general theory that a $(3, g)$-cage must have more than $n_0(3, g)$ vertices. The $(3, 7)$-cage is a graph with 24 vertices, and it is unique; details are given by Tutte (1966). There are numerous $(3, 9)$-cages; they have 58 vertices and the first one was found by Biggs and Hoare (1980). The fact that no smaller graph has degree 3 and girth 9 is the result of a computer search by B. McKay. There are three $(3, 10)$-cages; they have 70 vertices (O'Keefe and Wong 1980). The size of the $(3, 11)$-cage is as yet unknown. Since it is not a Moore graph it must have at least 96 vertices; the smallest known graph with degree 3 and girth 11 has 112 vertices.

23d *The Hoffman–Singleton graph*   The unique $(7,5)$-cage may be constructed by extending the graph described in **22f** as follows. Add 14 new vertices, called L, N, a, b, c, d, e, f, $1, 2, 3, 4, 5, 6$; join L to a, b, c, d, e, f and N; join N to $1, 2, 3, 4, 5, 6$ and L. Also, join the vertex denoted by (l, n) in **22f** to l and n. The automorphism group of this graph is the group of order 252 000 obtained from $PSU(3, 5^2)$ by adjoining the field automorphism of $GF(5^2)$ (Hoffman and Singleton 1960).

23e *Cages of girth 5 with $4 \leq k \leq 6$*   In these cases we know that a cage is not a Moore graph. There is a unique $(4, 5)$-cage with 19 vertices, due to Robertson (1964). There are several $(5, 5)$-cages having 30 vertices; see [BCN, p. 210]. There is a unique $(6, 5)$-cage (O'Keefe and Wong 1979); it has 40 vertices and it is the induced subgraph obtained by deleting the vertices of a Petersen graph from the Hoffman–Singleton graph.

23f *Cages of girth 6*   Recall (**4d**) that the *excess* of a k-regular graph with n vertices and girth g is $e = n - n_0(k, g)$. Biggs and Ito (1980) showed that, for small values of e, a k-regular graph with girth 6 and excess $e = 2(\eta - 1)$ is an η-fold covering of the incidence graph of a symmetric (v, k, η)-design.

When $\eta = 1$ such a design is a projective plane, and we have the generalized polygon graph as discussed above. When $\eta = 2$ such a design is called a *biplane*. In this case it can be shown that a necessary condition for the existence of a graph is that either k or $k - 2$ must be a perfect square (see Biggs 1981b). Such graphs with $k = 3$ and $k = 4$ do exist, but they they are not $(k, 6)$-cages, because for these values of k there is a generalized polygon graph. The first significant case is $k = 11$, because here it is now known that there is no projective plane, so the graph (if it exists) would be an $(11, 6)$-cage. There are several biplanes, but the existence of a 2-fold covering has not been settled.

When $\eta = 3$, coverings have been constructed for $k = 4, 7$ and 12. The case $k = 7$ is particularly important, because there is no projective plane or biplane in this case, and so the graph is a $(7, 6)$-cage (see O'Keefe and Wong (1981) and Ito (1981)). This is the last of the known cages.

23g *Families of graphs with large girth*   Graphs with small excess are very special, and we therefore adopt a wider definition of what is 'interesting' in this context. Let $\{\Gamma_r\}$ be a family of k-regular graphs such that Γ_r has n_r vertices and girth g_r. We say that the family has *large girth* if n_r and g_r both tend to infinity as $r \to \infty$, in such a way that

$$\lim_{r \to \infty} \frac{\log_{k-1} n_r}{g_r} \quad \text{is a finite constant } c.$$

It follows from the explicit form of $n_0(k, g)$ that c cannot be less than 0.5. For many years the existence of families with large girth was established only by non-constructive means; these arguments showed that there are families with $c = 1$. Weiss (1984) showed that, in the case $k = 3$, the family of sextet graphs $S(p)$ defined in **18b** has $c = 0.75$, and Lubotzky, Phillips and Sarnak (see **23h**) constructed families which attain the same value for infinitely many values of k. A simple construction for cubic graphs with large girth (but with $c > 1$) was given by Biggs (1987).

23h *The graphs of Lubotzky, Phillips and Sarnak*   Let p be a prime congruent to 1 modulo 4 and let H denote the set of integral quaternions $\alpha = (a_0, a_1, a_2, a_3)$. Define $\Lambda(2)$ to be the set of R-equivalence classes of elements α of H with $\alpha \equiv 1 \bmod 2$ and $\|\alpha\|$ a power of p, where $\alpha R \beta$ if $\pm p^r \alpha = p^s \beta$. Denote by S the set of elements of H satisfying $\|\alpha\| = p$, $\alpha \equiv 1 \bmod 2$, and $a_0 > 0$. There are $(p + 1)/2$ conjugate pairs $\{\alpha, \bar{\alpha}\}$ in S, and the Cayley graph of $\Lambda(2)$ with respect to S is the infinite $(p + 1)$-regular tree.

Now let q be another prime congruent to 1 modulo 4, such that $q > \sqrt{p}$ and $(p \mid q) = -1$. Denote by $\Lambda(2q)$ the normal subgroup of $\Lambda(2)$ consisting of those classes represented by α with a_1, a_2, a_3 divisible by $2q$. The Cayley graph of $S/\Lambda(2q)$ with respect to $\Lambda(2)/\Lambda(2q)$ is a bipartite $(p + 1)$-regular graph with $q(q^2 - 1)$ vertices and girth approximately $4 \log_p q$. For further details see Lubotzky, Phillips and Sarnak (1988), Biggs and Boshier (1990).

REFERENCES

Ádám, Á. (1967). Research problem 2-10, *JCT* **2**, 393.   [126]

Adel'son-Velskii, G.M., *et al.* (1969). Example of a graph without a transitive automorphism group [in Russian], *Soviet Math. Dokl.* **10**, 440–441.   [162]

Alon, N. and V.D. Milman (1985). λ_1, isoperimetric inequalities for graphs and superconcentrators, *JCT(B)* **38**, 73–88.   [22,58]

Alspach, B., D. Marušič, and L. Nowitz (1993). Constructing graphs which are 1/2-transitive, *J. Austral. Math. Soc.*, to appear.   [120]

Anthony, M.H.G. (1990). Computing chromatic polynomials, *Ars Combinatoria* **29C**, 216–220.   [111]

Aschbacher, M. (1971). The non-existence of rank three permutation groups of degree 3250 and subdegree 57, *J. Algebra* **19**, 538–540.   [188]

Austin, T.L. (1960). The enumeration of point labelled chromatic graphs and trees, *Canad. J. Math.* **12**, 535–545.   [41]

Babai, L. (1979). Spectra of Cayley graphs, *JCT(B)* **27**, 180–189. [128]

Babai, L. (1981). On the abstract group of automorphisms, *Combinatorics* (ed. H.N. Temperley, Cambridge U.P.), 1–40.   [117]

Baker, G.A. (1971). Linked-cluster expansion for the graph-vertex coloration problem, *JCT(B)* **10**, 217–231.   [81,86,93]

Bannai, E. and T. Ito (1973). On Moore graphs, *J. Fac. Sci. Univ. Tokyo Sec.* 1*A* **20**, 191–208.   [182]

Benson, C.T. (1966). Minimal regular graphs of girth eight and twelve, *Canad. J. Math.* **18**, 1091–1094.   [188]

Beraha, S. and J. Kahane (1979). Is the four-color conjecture almost false? *JCT(B)* **27**, 1–12.   [71]

Beraha, S., J. Kahane and N.J. Weiss (1980). Limits of chromatic zeros of some families of maps, *JCT(B)*, **28**, 52–65.   [71,72]

Biggs, N.L. (1970). *Finite Groups of Automorphisms* (Cambridge University Press).   [171]

Biggs, N.L. (1973a). Expansions of the chromatic polynomial, *Discrete Math.* **6**, 105–113.   [81]

Biggs, N.L. (1973b). Three remarkable graphs, *Canadian J. Math.* **25**, 397–411.   [103,111]

Biggs, N.L. (1973c). Perfect codes in graphs, *JCT(B)* **15**, 289–296.   [172]

Biggs, N.L. (1976). Automorphic graphs and the Krein condition, *Geom. Dedicata* **5**, 117–127.   [170]

Biggs, N.L. (1977a). Colouring square lattice graphs, *Bull. London Math. Soc.* **9**, 54–56.   [96]

Biggs, N.L. (1977b). *Interaction Models* (Cambridge U.P.). [80,87,110]

Biggs, N.L. (1978). On cluster expansions in graph theory and physics, *Quart. J. Math. (Oxford)* **29**, 159–173.   [81,87,90]

Biggs, N.L. (1979). Some odd graph theory, *Ann. New York Acad. Sci.* **319**, 71–81.   [58,137]

Biggs, N.L. (1980). Girth, valency and excess, *Lin. Alg. Appl.* **31**, 55–59.   [28]

Biggs, N.L. (1981a). Aspects of symmetry in graphs, *Algebraic Methods in Graph Theory* ed. L. Lovász and V. Sós (North-Holland) 27–35. [137]

Biggs, N.L. (1981b). Covering biplanes, *Theory and Applications of Graphs* ed. G. Chartrand (Wiley), 73–82.   [189]

Biggs, N.L. (1982a). A new 5-arc-transitive cubic graph, *J. Graph Theory* **6**, 447–451.   [154]

Biggs, N.L. (1982b). Constructing 5-arc-transitive cubic graphs, *J. London Math. Soc. (2)* **26**, 193–200.   [153]

Biggs, N.L. (1982c). Distance-regular graphs with diameter three, *Ann. Discrete Math.* **15**, 68–90.   [179]

Biggs, N.L. (1984a). Presentations for cubic graphs, *Computational Group Theory* ed. M.D. Atkinson (Academic Press), 57–63.   [146]

Biggs, N.L. (1984b). Homological coverings of graphs, *J. London Math. Soc. (2)* **30**, 1–14.   [154]

Biggs, N.L. (1987). Graphs with large girth, *Ars Combinatoria* **25C**, 73–80.   [190]

Biggs, N.L. and A.G. Boshier (1990). Note on the girth of Ramanujan graphs, *JCT(B)* **49**, 190–194.   [190]

Biggs, N.L., A.G. Boshier and J. Shawe-Taylor (1986). Cubic distance-regular graphs, *J. London Math. Soc. (2)* **33**, 385–394.   [163,171]

Biggs, N.L., G.R. Brightwell and D. Tsoubelis (1992). Theoretical and practical studies of a competitive learning process,   *Network* **3** 285–301.
[58]

Biggs, N.L., R.M. Damerell and D.A. Sands (1972). Recursive families of graphs, *JCT(B)* **12**, 123–131.   [42, 69, 70, 103]

Biggs, N.L. and M.J. Hoare (1980). A trivalent graph with 58 vertices and girth 9, *Discrete Math.* **30**, 299–301.   [189]

Biggs, N.L. and M.J. Hoare (1983). The sextet construction for cubic graphs, *Combinatorica* **3**, 153–165.   [144,152]

Biggs, N.L. and T. Ito (1980). Graphs with even girth and small excess, *Math. Proc. Cambridge Philos. Soc.* **88**, 1–10.   [189]

Biggs, N.L. and G.H.J. Meredith (1976). Approximations for chromatic polynomials, *JCT(B)* **20**, 5–19.   [96]

Biggs, N.L. and D.H. Smith (1971). On trivalent graphs, *Bull. London Math. Soc.* **3**, 155–158.   [171]

Birkhoff, G.D. (1912). A determinant formula for the number of ways of coloring a map, *Ann. Math.* **14**, 42–46.   [76]

Blokhuis, A. and A.E. Brouwer (1993). Locally $K_{3,3}$ or Petersen graphs, *Discrete Math.* **106/7**, 53–60.   [129]

Bollobás, B. (1985). *Random Graphs* (Academic Press).   [119]

Bondy, J.A. (1991). A graph reconstructor's manual, *Surveys in Combinatorics* ed. A.D. Keedwell (Cambridge U.P.), 221–252.   [51]

Bose, R.C. and D.M. Mesner (1959). Linear associative algebras, *Ann. Math. Statist.* **30**, 21–38.   [161]

Bouwer, I.Z. (1970). Vertex- and edge-transitive but not 1-transitive graphs, *Canad. Math. Bull.* **13**, 231–237.   [120]

Bouwer, I.Z. (1972). On edge- but not vertex-transitive regular graphs, *JCT(B)* **12**, 32–40.   [119]

Bouwer, I.Z. (1988). *The Foster Census* (Charles Babbage Research Centre, Winnipeg).   [147]

Brylawski, T. (1971). A combinatorial model for series-parallel networks, *Trans. Amer. Math. Soc.* **154**, 1–22.   [109]

Cameron, P.J. (1980). 6-transitive graphs, *JCT(B)* **28**, 168–179.   [162]

Cameron, P.J. (1982). There are only finitely many distance-transitive graphs of given valency greater than two, *Combinatorica* **2**, 9–13.   [163]

Cameron, P.J. (1991). Covers of graphs and EGQs, *Discrete Math.* **97**, 83–92.   [179]

194   References

Cameron, P.J., J.M. Goethals, J.J. Seidel, E.E. Shult (1976). Line graphs, root systems, and elliptic geometry, *J. Algebra* **43**, 305–327.   [18,22]

Cayley, A (1878). On theory of groups, *Proc. London Math. Soc.* **9**, 126–133.   [122]

Cayley, A. (1889). A theorem on trees, *Quart. J. Math.* **23**, 376–378. (*Collected papers*, **13**, 2628).   [39]

Chang, S. (1959). The uniqueness and nonuniqueness of the triangular association scheme, *Sci. Record* **3**, 604–613.   [21]

Chung, F-K. (1989). Diameters and eigenvalues, *J. Amer. Math. Soc.* **2**, 187–196.   [22]

Collatz, L. and U. Sinogowitz (1957) Spektren endlicher Grafen, *Abh. Math. Sem. Univ. Hamburg* **21**, 63–77   [12]

Conder, M. (1987). A new 5-arc-transitive cubic graph, *J. Graph Theory* **11**, 303–307.   [148]

Conder, M. and P. Lorimer (1989). Automorphism groups of symmetric graphs of valency 3, *JCT(B)* **47**, 60–72.   [128,146]

Coulson, C.A. and G.S. Rushbrooke (1940). Note on the method of molecular orbitals, *Proc. Camb. Philos. Soc.* 36, 193–200.   [11]

Coxeter, H.S.M. (1950). Self-dual configurations and regular graphs, *Bull. Amer. Math. Soc.* **59**, 413–455.   [153,154]

Crapo, H.H. (1967). A higher invariant for matroids, *JCT* **2**, 406–417.   [107]

Crapo, H.H. (1969). The Tutte polynomial, *Aeq. Math.* **3**, 211–229.   [102,104]

Cvetković, D.M. (1972). Chromatic number and the spectrum of a graph, *Publ. Inst. Math. (Beograd)* **14**, 25–38.   [57]

Cvetković, D.M. and P. Rowlinson (1988). Further properties of graph angles, *Scientia A: Math. Sci.* **1**, 41–51.   [51]

Damerell, R.M. (1973). On Moore graphs, *Proc. Cambridge Philos. Soc.* **74**, 227–236.   [160,182]

Delsarte, P. (1973). An algebraic approach to the association schemes of coding theory, *Philips Research Reports Suppl.* **10**.   [172]

Djoković, D.Ž. (1972). On regular graphs II, *JCT(B)* **12**, 252–259. [140]

Djoković, D.Ž. and G.L. Miller (1980). Regular groups of automorphisms of cubic graphs, *JCT(B)* **29**, 195–230.   [146]

Edge, W.L. (1963). A second note on the simple group of order 6048, *Proc. Cambridge Philos. Soc.* **59**, 1–9.   [188]

Egerváry, E. (1931). Matrizok kombinatorius tulajdonságairól, *Mat. Fiz. Lapok* **38**, 16–28   [35]

Elspas, B. and J.Turner (1970). Graphs with circulant adjacency matrices, *JCT* **9**, 297–307. [126]

Essam, J.W. (1971). Graph theory and statistical physics, *Discrete Math.* **1**, 83–112. [107]

Feit, W. and G. Higman (1964). The non-existence of certain generalized polygons, *J. Algebra* **1**, 114–131. [182]

Frucht, R. (1938). Herstellung von Graphen mit vorgegebener abstrakter Gruppe, *Compositio Math.* **6**, 239–250. [117]

Frucht, R. (1949). Graphs of degree three with a given abstract group, *Canad. J. Math.* **1**, 365–378. [117]

Frucht, R., J.E. Graver and M.E. Watkins (1971). The groups of the generalized Petersen graphs, *Proc. Cambridge Philos. Soc.* **70**, 211–218. [119]

Gardiner, A.D. (1973). Arc transitivity in graphs, *Quart. J. Math. Oxford* (2) **24**, 399–407. [137]

Gardiner, A.D. (1974). Antipodal covering graphs, *JCT(B)* **16**, 255–273. [179]

Gardiner, A.D. (1976). Homogeneous graphs, *JCT(B)* **20**, 94–102. [120]

Godsil, C.D. (1981). GRR's for solvable groups, *Algebraic Methods in Graph Theory* ed. L. Lovász and V. Sós (North-Holland), 221–239. [128]

Godsil, C.D. and A.D. Hensel (1992). Distance-regular covers of the complete graph, *JCT(B)* **56**, 205–238. [179]

Godsil, C.D. and B. Mohar (1988). Walk-generating functions and spectral measures of infinite graphs, *Lin. Alg. Appl.* **107**, 191–206. [13]

Goldschmidt, D. (1980). Automorphisms of trivalent graphs, *Ann. of Math.* **111**, 377–406. [143]

Grimmett, G.R. (1976). An upper bound for the number of spanning trees of a graph, *Discrete Math.* **16**, 323–324. [42]

Grimmett, G.R. (1978). The rank polynomials of large random lattices, *J. London Math. Soc.* **18**, 567–575. [94]

Grone, R. and R. Merris (1988). A bound for the complexity of a simple graph, *Discrete Math.* **69**, 97-99. [42]

Gross, J.L. (1974). Voltage graphs, *Discrete Math.* **9**, 239–246. [149]

Grötschel, M., L. Lovász, and A. Schrijver (1988). *Geometric Algorithms and Combinatorial Optimization* (Springer, New York). [35]

Haemers, W. (1979). *Eigenvalue Techniques in Design and Graph Theory* (Math. Centrum, Amsterdam). [57]

Haggard, G. (1976). The chromatic polynomial of the dodecahedron, Technical Report No. 11 (University of Maine). [69]

Hall, J.I. (1980). Locally Petersen graphs, *J. Graph Theory*, **4**, 173–187.
[129]

Hammersley, J. (1981). The friendship theorem and the love problem, *Surveys in Combinatorics* ed. E.K. Lloyd (Cambridge University Press), 31–54.
[171]

Harary, F. (1962). The determinant of the adjacency matrix of a graph, *SIAM Review* **4**, 202–210.
[44]

Heller, I. and C.B. Tompkins. An extension of a theorem of Dantzig's, *Linear Inequalities and Related Systems* ed. H.W. Kuhn and A.W. Tucker (Princeton U.P.).
[34,35]

Heron, A. (1972). Matroid polynomials, *Combinatorics* ed. D.J.A. Welsh and D.R. Woodall (Institute of Mathematics and its Applications) 164–202.
[109]

Hetzel, D. (1976). Über reguläre graphische Darstellung von auflösbaren gruppen, *Diplomarbeit, Technische Universität Berlin.*
[128]

Higman, D.G. (1967). Intersection matrices for finite permutation groups, *J. Algebra* **6**, 22–42.
[161]

Higman, D.G. and C.C. Sims (1968). A simple group of order 44,352,000, *Math. Zeitschr.* **105**, 110–113.
[161]

Hoffman, A.J. (1960). On the exceptional case in the characterization of the arcs of a complete graph, *IBM J. Res. Dev.* **4**, 487–496.
[21]

Hoffman, A.J. (1963). On the polynomial of a graph, *Amer. Math. Monthly* **70**, 30–36.
[15]

Hoffman, A.J. (1969). The eigenvalues of the adjacency matrix of a graph, *Combinatorial Mathematics and its Applications* (Univ. of N. Carolina Press, Chapel Hill), 578–584.
[17]

Hoffman, A.J. (1970). On eigenvalues and colorings of graphs, *Graph Theory and its Applications* (Academic Press, New York), 79–92.
[57]

Hoffman, A.J. and J.B. Kruskal (1956). Integer boundary points of convex polyhedra, *Linear Inequalities and Related Systems* ed. H.W. Kuhn and A.W. Tucker (Princeton U.P.), 223–246.
[35]

Hoffman, A.J. and R.R. Singleton (1960). On Moore graphs with diameters 2 and 3, *IBM J. Res. Dev.* **4**, 497–504.
[182,188,189]

Holt, D. (1981). A graph which is edge-transitive but not arc-transitive, *J. Graph Theory* **5**, 201–204.
[120]

Horton, J.D. and I.Z. Bouwer (1991). Symmetric Y-graphs and H-graphs, *JCT(B)* **53**, 114–129.
[147]

Ito, T. (1981). On a graph of O'Keefe and Wong, *J. Graph Theory* **5**, 87–94.
[190]

Ito, T. (1982). Bipartite distance-regular graphs of valency three, *Lin. Alg. Appl.* **46**, 195–213.
[153]

Ivanov, A.A. (1983). Bounding the diameter of a distance-regular graph, *Soviet Math. Dokl.* **28**, 149–152. [161]

Izbicki, H. (1960). Reguläre Graphen beliebigen Grades mit vorgegebenen Eigenschaften, *Monatshefte für Math.* **64**, 15–21. [121]

Jaeger, F. (1988). On Tutte polynomials and cycles of planar graphs, *JCT(B)* **44**, 129–146. [104]

Jaeger, F. (1991). Even subgraph expansions for the flow polynomial of cubic plane maps, *JCT(B)* **52**, 259–273. [110]

Jaeger, F., D.L. Vertigan and D.J.A. Welsh (1990). The computational complexity of the Jones and Tutte polynomials, *Math. Proc. Camb. Philos. Soc.* **108**, 35–53. [105,111]

Jordan, C. (1869). Sur les assemblages des lignes, *J. f. d. Reine u. Angew. Math.* **70**, 185–190. [119]

Kelly, P.J. (1957). A congruence theorem for trees. *Pacific J. Math.* **7**, 961–968. [50]

Kim, D. and I.G. Enting (1979). The limit of chromatic polynomials, *JCT(B)* **26**, 327–336. [96]

Kirchhoff, G. (1847). Über die Auflösung der Gleichungen, auf welche man bei der Untersuchung der linearen Verteilung galvanischer Ströme gefürht wird, *Ann. Phys. Chem.* **72**, 497–508. [31,36]

Kocay, W.L. (1981). On reconstructing spanning subgraphs, *Ars Combinatoria* **11**, 301–313. [50]

Las Vergnas, M. (1978). Eulerian circuits of 4-valent graphs imbedded in surfaces, *Algebraic Methods in Graph Theory* ed. L. Lovász and V. Sós (North-Holland, Amsterdam), 451–477. [104]

Las Vergnas, M. (1988). On the evaluation at (3,3) of the Tutte polynomial of a graph, *JCT(B)* **45**, 367–372. [104]

Li, N-Z., and E.G. Whitehead (1992). The chromatic uniqueness of W_{10}, *Discrete Math.* **104**, 197–199. [69]

Lieb, E.H. (1967). Exact solution of the problem of the entropy of two-dimensional ice, *Phys. Rev. Letters* **18**, 692–694. [95]

Lorimer, P. (1989). The construction of Tutte's 8-cage and the Conder graph, *J. Graph Theory* **13**, 553–557. [148]

Lovász, L. (1979). On the Shannon capacity of a graph, *IEEE Trans. Inform. Theory* **25**, 1–7. [51]

Lubotzky, A., R. Phillips and P. Sarnak (1988). Ramanujan graphs, *Combinatorica* **8**, 261–277. [190]

Meredith, G.H.J. (1976). Triple-transitive graphs, *J. London Math. Soc.* *(2)* **13**, 249–257. [162]

Mohar, B. (1991). Eigenvalues, diameter, and mean distance in graphs, *Graphs Comb.* **7** 53–64. [22]

References

Moon, J. (1967). Enumerating labelled trees, *Graph Theory and Theoretical Physics* (Academic Press, New York), 261–272. [41]

Moon, J. (1970). *Counting labelled trees* (Canadian Math. Congress, Montreal). [38]

Motzkin, T., and E.G. Straus (1965). Maxima for graphs and a new proof of a theorem of Turán, *Canad. J. Math.* **17**, 533–540. [59]

Mowshowitz, A. (1969). The group of a graph whose adjacency matrix has all distinct eigenvalues, *Proof Techniques in Graph Theory* (Academic Press, New York), 109–110. [117]

Nagle, J.F. (1971). A new subgraph expansion for obtaining coloring polynomials for graphs, *JCT(B)* **10**, 42–59. [80]

Nerode, A. and H. Shank (1961). An algebraic proof of Kirchhoff's network theorem, *Amer. Math. Monthly* **68**, 244–247. [36]

O'Keefe, M. and P.K. Wong (1979). A smallest graph of girth 5 and valency 6, *JCT(B)* **26**, 145–149. [189]

O'Keefe, M. and P.K. Wong (1980). A smallest graph of girth 10 and valency 3, *JCT(B)* **29**, 91–105. [189]

O'Keefe, M. and P.K. Wong (1981). The smallest graph of girth 6 and valency 7, *J. Graph Theory* **5**, 79–85. [190]

Parsons, T.D. (1980). Circulant graph imbeddings, *JCT(B)* **29**, 310–320. [126]

Petersdorf, M. and H. Sachs (1969). Spektrum und Automorphismengruppe eines Graphen, *Combinatorial Theory and its Applications, III* (North-Holland, Amsterdam), 891–907. [117,125]

Poincaré, H. (1901). Second complément à l'analysis situs, *Proc. London Math. Soc.* **32**, 277–308. [32]

Ray, N. (1992). Tutte algebras of graphs and formal group theory, *Proc. London Math. Soc.* **65**, 23–45. [88]

Ray, N. and C. Wright (1992). Umbral interpolation and the addition/contraction tree for graphs, *Discrete Math.* **103**, 67–74. [72]

Read, R.C. (1968). An introduction to chromatic polynomials, *J. Comb. Theory* **4**, 52–71. [69,108]

Read, R.C. (1990). Recent advances in chromatic polynomial theory, *Proc. 5th Carib. Conf. on Graph Theory and Computing.* [71]

Robertson, N. (1964). The smallest graph of girth 5 and valency 4, *Bull. Amer. Math. Soc.* **70**, 824–825. [189]

Ronse, C. (1978). On homogeneous graphs, *J. London Math. Soc. (2)* **17**, 375–379. [120]

Rowlinson, P. (1987). A deletion–contraction algorithm for the characteristic polynomial of a multigraph, *Proc. Roy. Soc. Edin.* **105A**, 153–166. [11]

Sachs, H. (1962). Über selbstkomplementäre Graphen, *Publ. Math. Debrecen* **9**, 270–288. [20]

Sachs, H. (1963). Regular graphs with given girth and restricted circuits, *J. London Math. Soc.* **38**, 423–429. [180]

Sachs, H. (1964). Beziehungen zwischen den in einem Graphen enthaltenen Kreisen und seinem charakterischen Polynom, *Publ. Math. Debrecen.* **11**, 119–134. [49]

Sachs, H. (1967). Über Teiler, Faktoren und characterische Polynome von Graphen II, *Wiss. Z. Techn. Hosch. Ilmenau* **13**, 405–412. [19]

Schmitt, W. (1993). Hopf algebra methods in graph theory, *Pure Appl. Algebra*, to appear. [88]

Schur, I. (1933). Zur Theorie der einfach transitiven Permutationsgruppen, *S.B. Preuss. Akad. Wiss.*, 598–623. [161]

Schwenk, A.J. (1973). Almost all trees are cospectral, *New Directions in the Theory of Graphs* (Academic Press, New York). [49]

Scott, L.L. (1973). A condition on Higman's parameters, *Notices Amer. Math. Soc.* **20**, A97 (701-20-45). [170]

Sedlacek, J. (1970). On the skeletons of a graph or digraph, *Combinatorial Structures and Applications* (Gordon and Breach, New York), 387–391. [42]

Seidel, J.J. (1968). Strongly regular graphs with $(-1, 1, 0)$ adjacency matrix having eigenvalue 3, *Lin. Alg. Appl.* **1**, 281–298. [18,21]

Seidel, J.J. (1979). Strongly regular graphs, *Surveys in Combinatorics* ed. B. Bollobás (Cambridge U.P.), 157–180. [171]

Seymour, P.D. (1981). Nowhere-zero 6-flows. *JCT(B)* **30**, 130–135. [30]

Sheehan, J. (1974). Smoothly embeddable subgraphs, *J. London Math. Soc. (2)* **9**, 212–218. [120]

Shrikhande, S.S. (1959). The uniqueness of the L_2 association scheme, *Ann. Math. Stat.* **30**, 781–798. [21]

Sims, C.C.(1967).Graphs and finite permutation groups, *Math. Zeitschr.* **95**, 76–86. [141]

Singleton, R.R. (1966). On minimal graphs of maximum even girth, *J. Comb. Theory* **1**, 306–332. [188]

Smith, D.H. (1971). Primitive and imprimitive graphs, *Quart. J. Math. Oxford (2)* **22**, 551–557. [177,178]

Smith, J.H. (1970). Some properties of the spectrum of a graph, *Combinatorial Structures and their Applications* ed. R. Guy *et al.* (Gordon and Breach, New York), 403–406. [43]

Stanley, R.P. (1973). Acyclic orientations of graphs, *Discrete Math.* **5**, 171–178. [70]

Temperley, H.N.V. (1964). On the mutual cancellation of cluster integrals in Mayer's fugacity series, *Proc. Phys. Soc.* **83**, 3–16. [39]

Thistlethwaite, M.B. (1987). A spanning tree expansion of the Jones polynomial, *Topology* **26**, 297–309. [105]

Thomassen, C. (1990). Resistances and currents in infinite electrical networks, *JCT(B)* **49**, 87–102. [36]

Tutte, W.T. (1947a). A family of cubical graphs, *Proc. Cambridge Philos. Soc.* **45**, 459–474. [130,141,143]

Tutte, W.T. (1947b). A ring in graph theory, *Proc. Cambridge Philos. Soc.* **43**, 26–40. [79]

Tutte, W.T. (1954). A contribution to the theory of chromatic polynomials, *Canad. J. Math.* **6**, 80–91. [30,102,110]

Tutte, W.T. (1956). A class of Abelian groups, *Canad. J. Math.* **8**, 13–28. [30]

Tutte, W.T. (1966). *Connectivity in Graphs* (Toronto University Press). [120,132, 189]

Tutte,W.T.(1967). On dichromatic polynomials, *JCT* **2**, 301–320. [81,86,90,96]

Tutte, W.T. (1970). On chromatic polynomials and the golden ratio, *JCT*, **9** 289–296. [71]

Tutte, W.T. (1974). Codichromatic graphs, *JCT(B)* **16**, 168–174. [79]

Tutte, W.T. (1979). All the king's horses – a guide to reconstruction, *Graph Theory and Related Topics* ed. J.A. Bondy and U.S.R. Murty (Academic Press), 15–33. [57]

Tutte, W.T. (1984). *Graph Theory* (Addison-Wesley). [72]

Vijayan, K.S. (1972). Association schemes and Moore graphs, *Notices Amer. Math. Soc.* **19**, A-685. [182]

Weiss, A. (1984). Girths of bipartite sextet graphs, *Combinatorica* **4**, 241–245. [190]

Weiss, R. (1974). Über s-reguläre Graphen, *JCT(B)* **16**, 229-233. [141]

Weiss, R. (1983). The non-existence of 8-transitive graphs, *Combinatorica* **1**, 309–311. [137]

Weiss, R. (1985) On distance-transitive graphs, *Bull. London Math. Soc.* **17**, 253–256. [163]

Whitney, H. (1932a). A logical expansion in mathematics, *Bull. Amer. Math. Soc.* **38**, 572–579. [76,77]

Whitney, H. (1932b). The coloring of graphs, *Ann. Math.* **33**, 688–718. [69,81,86]

Whitney, H. (1932c). Congruent graphs and the connectivity of graphs, *Amer. J. Math.* **54**, 150–168. [120]

Wielandt, H. (1964). *Finite Permutation Groups* (Academic Press).
[161,172]

Wilf, H.S. (1967). The eigenvalues of a graph and its chromatic number, *J. London Math. Soc.* **42**, 330–332. [55]

Wilf, H.S. (1986). Spectral bounds for the clique and independence numbers of graphs, *JCT(B)* **40**, 113–117. [59]

Woodall, D.R. (1992). A zero-free interval for chromatic polynomials, *Discrete Math.* **101**, 333–341. [72]

Yetter, D.N. (1990). On graph invariants given by linear recurrence relations, *JCT(B)* **48**, 6–18. [72]

Yuan, Hong (1988). A bound on the spectral radius of graphs, *Lin. Alg. Appl.* **108**, 135–139. [13]

Index

acyclic orientation 70
adjacent 7
adjacency algebra 9
adjacency matrix 7
almost-complete 43
alternating knot 105
angles 51
antipodal 177
antipodal r-fold covering 178
augmentation 29
automorphic 178
automorphism 115
automorphism group 115

bicentroid 119
bigrading 97
bipartite 11
biplane 189
block 81
block system 173
broken cycle 77
Brooks's theorem 55
buckminsterfullerene 127

cage 181, 188, 189
Cayley graph 123
centroid 119
characteristic polynomial 8
chromatically unique 69
chromatic invariant 107
chromatic number 52
chromatic polynomial 63
chromatic root 71
circulant graph 16, 126
circulant matrix 16
closed walk 12
coboundary mapping 28
cocktail-party graph 17, 68
colour-class 52
colour-partition 52
compatible 150
complete bipartite graph 21

complete graph 8
complete matching 50
complete multipartite graph 41
conductance 34
cone 66
confluence 72
conforms 30
conjugate Bell polynomials 72
connected 10
contracting 64
Conway's presentations 145
co-rank 25, 97
coset graph 128
cospectral graphs 12, 49
cover 50
covering graph 149
cube 43, 69, 140, 157, 161, 169
cubic graph 138
current 34
cut 26
cut-orientation 26
cut-subspace 26
cut-vertex 67
cycle 25
cycle graph 17, 65
cycle-orientation 25
cycle-subspace 26

degree 4
deletion-contraction 65, 72
density 94
derived graph 178
Desargues graph 148, 153
diameter 10
dihedral group 126
distance 10
distance matrices 13, 159
distance-regular 13, 159
distance-transitive 118, 155
dodecahedron 69, 178
double pyramid 68

double-transitivity 118
dual 29,43

edge 3
edge space 23
edge-transitive 115, 118, 120
effective resistance 36
eigenvalue 8
electrical network 34
elementary 44
ends 4
equipartition 58
even subgraph 110
excess 28, 189
expansion 147
external activity 99
externally active 99

feasible array 168
flow 29
flow polynomial 110
forest 47
Foster's census 147
friendship theorem 171

generalized d-gon 187
generalized line graph 21
generalized polygon graph 181
general graph 3
girth 28, 76, 131, 180
graph 4
graphical regular representation
    124, 128
graph types 87

Hamiltonian cycle 50
Hamming graph 169
Heawood graph 148, 154, 163
Hoffman-Singleton graph 189
homeomorphic 79, 108
homogeneous 120
homological covering 154
Hopf algebra 88
hyperoctahedral graph 17

icosahedral group 127
icosahedron 69, 178
imprimitive 177

incidence mapping 24, 29
incidence matrix 24
independent 98
indicator function 74
induced subgraph 4
interaction model 80
internal activity 99
internally active 99
intersection array 157, 159
intersection matrix 165
intersection numbers 156
Ising model 80
isoperimetric number 28, 58
isthmus 30

join 66
Jones polynomial 105

K-chain 149
Kelly's lemma 50
Kirchhoff's laws 34
Kocay's lemma 50
Krein parameters 170

labelled tree 104
ladder 69, 126
Laplacian matrix 27
Laplacian spectrum 29, 40
line graph 17, 120
logarithmic transform 82
loop 3

medial graph 104
minimal support 29
Möbius ladder 20, 42, 69, 110
modified rank polynomial 101
modular flow 30
Moore graph 181
Motzkin-Straus formula 59

negative end 24
nowhere-zero 30

octahedron 43
odd graphs 20, 58, 137, 161, 170
orbit 115
orientation 24

Paley graph 129

Pappus graph 148, 154
partial geometry 162
partition function 80
path graph 11
perfect code 22, 171
permutation character 172
permutation matrix 116
Petersen graph 20, 95, 103, 133
planar 29
positive end 24
potential 36
Potts model 80
power 36
primitive 30, 173
principal minors 8
projective plane 163
proper 90
pyramid 68

quasi-separable 67
quasi-separation 67

rank 25
rank matrix 73
rank polynomial 73
Rayleigh quotient 54
Rayleigh's monotonicity law 37
reconstructible 50, 91
reconstruction conjecture 50
recursive family 70, 103
regular graph 14
regular action 122
resonant model 80
rewriting rules 72
root systems 22
r-ply transitive 162

semi-direct product 150
separable 67
separation 67
series-parallel 109
sextet graph 145
Shannon capacity 51
sides 149
simple eigenvalues 116, 125
spanning elementary subgraph 44
spanning tree 31
spectral decomposition 13

spectrum 8
sporadic groups 172
square lattice 96
stabilizer 122
stabilizer sequence 133, 137, 147
standard bases 24
star graph 49
star types 87
strict graph 4
strongly regular graph 16, 20, 159, 171
subdividing 79
subgraph 4
successor 132
support 29
suspension 66
symmetric 118, 126
symmetric cycle 137
symmetric design 163
symmetric group 118, 148

t-arc 130
tetrahedral group 127
thermodynamic limit 94
theta graph 86
Thomson's principle 36
topological invariant 79
totally unimodular 34
tree 47, 49, 65, 119
tree-number 38
triangle graph 19, 169
tridiagonal 165
t-transitive 131
Turan's Theorem 59
Tutte polynomial 97, 100

umbral chromatic polynomial 72
unimodal conjecture 108

vertex 3
vertex-colouring 52
vertex space 23
vertex-stabilizer 122, 127
vertex-transitive 115, 120, 125
V-function 79
voltage 34

walk 9

walk-generating function 13
walk-generating matrix 12

weakly homogeneous 120
wheel 68